# THE TOOLS OF EMPIRE

# THE TOOLS OF EMPIRE
### Technology and
### European Imperialism
### in the Nineteenth Century

DANIEL R. HEADRICK

New York   Oxford
OXFORD UNIVERSITY PRESS
1981

Copyright © 1981 by Oxford University Press, Inc.

Library of Congress Cataloging in Publication Data

Headrick, Daniel R
The tools of empire.

Bibliography: p.
Includes index
1. Imperialism—History. 2. Technology—
History. I. Title.
JC359.H4  303.4′83′09034  80-18099
ISBN 0-19-502831-7
ISBN 0-19-502832-5 (pbk.)

Printed in the United States of America

# Preface

For the inspiration, encouragement, and assistance that have contributed so much to this book, my debts run deep—to Professors William H. McNeill and Ralph Austen of the University of Chicago; to Zohar Ben Asher; to Gérard Fabres; to David Northrup; and to the librarians of so many superb institutions: in Chicago the Regenstein Library; in Paris the Archives Nationales Section Outre-Mer, Bibliothèque Nationale, and Maison des Sciences de l'Homme; in London the India Office Library and Records, British Library, National Maritime Museum, and School of Oriental and African Studies; in Brazzaville (Congo) the Organisation de la Recherche Scientifique et Technique Outre-Mer and the Université Marien Ngouabi. For permission to use material from The Carl H. Pforzheimer Library in New York, I thank the Carl and Lily Pforzheimer Foundation, Inc.

This book is dedicated to my father and the memory of my mother, who nurtured my bookishness; to my children, Isabelle, Juliet, and Matthew, whose happy company cheered my working hours; and to Rita, my wife, not for her typing but for her thoughts.

*Chicago*                                                        D.R.H.
*January 1981*

v

# Contents

# THE TOOLS OF EMPIRE

# Technology, Imperialism, and History

Among the many important events of the nineteenth century, two were of momentous consequence for the entire world. One was the progress and power of industrial technology; the other was the domination and exploitation of Africa and much of Asia by Europeans. Historians have carefully described and analyzed these two phenomena, but separately, as though they had little bearing on each other. It is the aim of this book to trace the connections between these great events.

The European imperialism of the nineteenth century—sometimes called the "new" imperialism—differed from its precursors in two respects: its extent and its legacy. In the year 1800 Europeans occupied or controlled thirty-five percent of the land surface of the world; by 1878 this figure had risen to sixty-seven percent, and by 1914 over eighty-four percent of the world's land area was European-dominated.[1] The British Empire alone, already formidable in 1800 with a land area of 1.5 million square miles and a population of twenty million, increased its land area sevenfold and its population twentyfold in the following hundred years.[2]

Its legacy is harder to quantify. In today's Asia and Africa,

European political and religious ideals barely survive as tenuous mementos of a defunct imperial age—modern equivalents of the Mosque of Cordoba or Hadrian's Wall. The real triumph of European civilization has been that of vaccines and napalm, of ships and aircraft, of electricity and radio, of plastics and printing presses; in short, it has been a triumph of technology, not ideology. Western industrial technology has transformed the world more than any leader, religion, revolution, or war. Nowadays only a handful of people in the most remote corners of the earth survive with their lives unaltered by industrial products. The conquest of the non-Western world by Western industrial technology still proceeds unabated.

This conquest began in the nineteenth century and was woven into the expansion of European empires. The connections between technology and imperialism must be approached from both sides: from the history of technology as well as from that of imperialism. The history of technology ranks as one of the more popular forms of literature. Bookstores with only a handful of biographies and national histories will offer whole shelves of books on the history of guns, antique furniture, vintage cars, old-time locomotives, and Nazi warplanes. Most of these are hardware histories, compilations of pictures and facts about objects divorced from the context of their time. The social history of technology, in contrast, aims at understanding the causes, the development, and the consequences of technological phenomena. Social historians of technology generally begin with a given technology and examine it in this light. Examples abound: What was the impact of the cotton industry on British labor during the Industrial Revolution? How did firearms change warfare during the late Middle Ages? How did railroads contribute to the winning of the American West? Reversing the question, however, can also shed light on the historical process. Given a particular historical phenomenon— for example, the new imperialism—how did technological forces shape its development? This is the question that historians of imperialism have neglected to answer, and that we must now confront.

The search for the causes of nineteenth-century imperialism has spawned one of the liveliest debates in modern history. Historians have offered a wealth of explanations for this dramatic expansion of European power. Some have emphasized such political motives as international rivalries, naval strategy, the instability of imperial frontiers, the diversion of popular attention from domestic problems, or the influence of pressure groups on political decision makers. Others, following in the footsteps of the English economist J. A. Hobson, have stressed economic motives: the need for raw materials, secure markets, or investment opportunities.[3]

At first sight, these points of view seem to differ markedly. Yet, we are struck by a common underlying element. Participants in the debate agree that the crucial factor in the new imperialism was the motivation of the imperialists. What made nineteenth-century politicians, explorers, traders, missionaries, and soldiers want to extend the influence of Europe to hitherto untouched lands? Behind this question lies the tacit assumption that once Europeans wanted to spread their influence, they could readily do so, for they had the means close at hand.

Rather than analyze in detail the debate on the new imperialism, let us consider a few recent and important contributions. These can be divided into three broad categories: those that ignore the role of technology, those that disparage it, and those that gloss over it lightly. In the first category is *Africa and the Victorians* by Ronald Robinson and John Gallagher. The authors consider the conquest of Africa in the last decades of the nineteenth century:

> Why, after centuries of neglect, the British and other European governments should have scrambled to appropriate nine tenths of the African continent within sixteen years, is an old problem,

still awaiting an answer. . . . What were the causes and incentives? Which of them were merely contributory and which decisive? . . . A first task in analyzing the late-Victorians' share in the partition is to understand the motives of the ministers who directed it. . . .[4]

Having asked these questions, the authors find that indeed the scramble for Africa can be explained by the intentions and hesitations of the statesmen, the "official mind" of the European powers.

A similar perspective pervades Henri Brunschwig's *French Colonialism 1871–1914, Myths and Realities*.[5] In this book the author argues that France acquired an empire mainly for psychological reasons: wounded pride following the Franco-Prussian War, and the desire to regain status and prestige among the great powers. A less Eurocentric theory is advanced by D. K. Fieldhouse in *Economics and Empire 1830–1914*. Imperialism, defined as military and political conquest, was the consequence of instability generated on the frontiers of empire by advancing parties of traders, missionaries, and other Europeans coming into conflict with indigenous societies: ". . . imperialism may be seen as a classic case of the metropolitan dog being wagged by its colonial tail," he declares, or again, ". . . Europe was pulled into imperialism by the magnetic force of the periphery."[6] After presenting myriad instances of this phenomenon, however, Fieldhouse is left with the tantalizing question:

Is 'imperialism' merely shorthand for an agglometration of causally unrelated events which happened to occur at largely the same time in different parts of the world? If so, why did the critical period of imperialism happen to occur in these thirty years after 1880?

These multiple crises and their timing were merely symptoms of a profound change in the pathology of international relationships. The world crisis was real and a solution had to be found. By about 1880 there was a profound disequilibrium between Europe and most parts of the less-developed world. Never had one continent possessed so immense a power advantage over the others or been in such close contact with them.[7]

Imperialism, then, was the sum of many little imperialisms tied together by their timing. And their timing was the product of "a profound change in the pathology of international relationships," "a profound disequilibrium," "a power advantage." Here, on the very brink of offering a concrete explanation for the new imperialism as a unified movement, Fieldhouse backs off, leaving us with hazy, mysterious forces.

In the works cited above, the authors disregard the technological factor in imperialism. Others mention it, but only to reject it. This is the position taken by both Hans-Ulrich Wehler and Rondo Cameron.

In his book *Imperialismus,* Wehler asserts:

> . . . technological progress as such did not cause imperialistic expansion directly, let alone automatically, but contributed as the impetus in other areas. Imperialism resulted, in a way, from the socio-political inability, within the political framework, to cope with the economic results of permanent technological innovations and their social consequences.[8]

In a later work, Wehler is even more categorical:

> If one points to technological progress as the main factor of expansion, thereby defining imperialism as a sort of unavoidable "natural" consequence of technological innovations, one is led astray too. There is no direct causal relationship between these innovations and imperialism.[9]

Rondo Cameron reached a similar conclusion in an article entitled "Imperialism and Technology," which appeared in a general history of technology:

> It is sometimes asserted that the rapid progress of Western technology in the 19th century was a major determinant of the imperialist drive. . . . Western superiority in ships, navigational techniques, and firearms was a fact of long standing, however. It

cannot be used to explain the burst of expansion at the end of the 19th century, after almost a century during which Europeans showed little interest in overseas expansion.[10]

The rejection of the technological factor by Wehler and Cameron, and its disregard by Robinson, Gallagher, Brunschwig, and Fieldhouse are not representative of the literature as a whole. The majority of current works on imperialism concede the importance of the technological factor, paying it lip-service before hurrying on to something else. An excellent example is the recent *African History* by Philip Curtin and others. The authors recognize that, as a result of advances in medicine and pharmacology, iron and steel, and firearms, "conquest in Africa was not only far cheaper than it had ever been in the past; it was also far cheaper in lives and money than equivalent operations would ever be again." Yet to these "technological factors," they devote only three pages.[11]

The conclusion is inescapable; at the present stage of the debate, historians give technological factors very low priority among the causes of the new imperialism. Such a curt dismissal of the role of technology in nineteenth-century imperialism stands in striking contrast to the central role assigned to technological change—better known as the Industrial Revolution—in the histories of European societies and economies of the very same period. It contrasts even more with the careful attention that historians of the early modern period have devoted to the technological aspects of the oceanic discoveries and of the exploration and conquest of the Americas.[12]

One reason for the disregard of technological factors lies in the leading-sectors model of the Industrial Revolution. This widely accepted explanation concentrates on the role of the most innovative and fastest-expanding industries—textiles, railroads, mining, and metallurgy—which exerted strong multiplier effects on the rest of the economy. It is quite reasonable for someone to consider these leading sectors and conclude that they became important in the non-Western world only late in the colonial period, and not in the earlier periods of penetration and conquest.

If the more dramatic aspects of the Industrial Revolution had only a marginal impact on imperial expansion, it does not follow that technology in general was inconsequential. In order to discover which inventions were important on the frontiers of empire, we must look at Africa and Asia as well as Europe, and at indigenous technologies and natural obstacles as well as at the technologies of the imperialists.

A more fundamental reason for misunderstanding the role played by technology lies in the very concept of causality used by historians. Almost all historians nowadays view imperialism as the result of many causes; their interpretations differ in the weights they attach to each cause. The problem with this way of thinking is that any attempt to emphasize the role of one cause automatically reduces the importance of the others, thereby coming into conflict with other interpretations. The debate on the new imperialism is essentially the result of conflicts in the ordering of causes. To defend the importance of a new factor is therefore to run head-on into other interpretations. And to advance the claims of technology—which many still associate with the concept of matter over mind—seems at first to defy an axiom of Western historiography: that history results from the interactions of human decisions.

This dilemma is much relieved if we divide causes into motives and means. A complex process like imperialism results from both appropriate motives and adequate means. If the motives are too weak—as they were in the case of the Chinese expeditions to the Indian Ocean in the 1430s—or if the means are inadequate—as in the Italian invasion of Ethiopia in the 1890s—then the imperialist venture aborts. Both types of causes are equally important, and by focusing on one we in no way discredit the other.

A model of causality in which the technical means are separate from the motives does not imply that the two are unrelated. On the contrary, the appearance of a new technology can trigger or reinforce a motive by making the desired end possible or acceptably inexpensive. For example, quinine prophylaxis allowed Europeans to survive in tropical Africa. Conversely, a motive can occasion a search for appropriate means, as when the American occupation of Cuba brought about the investigation of the causes of yellow fever. The many instances in which we will encounter both types of relationships will serve as reminders that we must steer between two dangerous determinisms: the technological—"what can be done will be done"—and the psychological—"where there's a will there's a way."

If we accept the equal necessity of both motives and means, then the new imperialism could have resulted from any of three possible scenarios: Adequate means were available, but new motives triggered the event; sufficient motives existed, but new means came into play, thus leading to the event; or, finally, both the motives and the means changed, and both caused the event.

The first scenario—which Cameron sums up with the words "Western superiority . . . was a fact of long standing"—has formed the basis of the debate up until recently. Yet Western superiority, insofar as it existed, does not suffice to explain European conquests in Asia and Africa. The new imperialism was not the result of mere superiority, but of the unleashing of overwhelming force at minimal costs. Technological changes affected the timing and location of the European conquests. They determined the economic relations of colonialism. And they paved the way for the astonishing reversal of the world balance that we are currently witnessing.

If the first scenario overemphasizes motivations, the second one gives more credit to technological factors than the evidence warrants. Europeans were not always equally interested in Asia and Africa, and historians have rightly stressed the growing demand for colonies in the late nineteenth century.

It is the third of our scenarios, in which both means and motivations changed and interacted, that best reflects the realities of the European conquest and colonization of the eastern hemisphere during the nineteenth century. It is the purpose of this book to argue this third scenario, by analyzing the technological changes that made imperialism happen, both as they enabled motives to produce events, and as they enhanced the motives themselves.

The goal and result of imperialism—one which was in fact achieved in most territories before decolonization—was the creation of colonies politically submissive and economically profitable to their European metropoles. The economic networks that were established, and the technologies that entered into the development and exploitation of colonial plantations, farms, mines, and forests, are a complex subject that we must leave for another time.

This book concentrates on an earlier period, that of imperial expansion. The imperialism of Europe in Asia and Africa involved a number of stages before the goal of pacified colonies eventually was reached. Though these occurred at different times and in different ways depending on the region, we can classify these stages roughly as follows. The initial stage was that of penetration and exploration by the first European travelers. Then came the conquest of the indigenous people and the imposition of European rule on them. Finally, before the colony could become valuable as an adjunct to a European economy, a communications and transportation network had to be forged.

From the technological point of view, each of these stages involved hundreds of diverse products and processes, from pith-helmets to battleships. In this book I will concentrate on those that played the most crucial role, either by making imperialism possible where it was otherwise unlikely, or by making it suitably cost-effective in the eyes of budget-minded governments. In the penetration phase, steamers and the prophylactic use of quinine were the key technologies. The second phase—that of conquest—depended heavily on rapid-firing rifles and machine guns. In the phase of consolidation, the links that tied the colonies to Europe and promoted their economic exploitation included steamship lines, the Suez Canal, the submarine telegraph cables, and the colonial railroads. These technological factors are the subject of this work.

The effects of technological change were experienced almost everywhere in the nineteenth century, but they were felt much more strongly in some parts of the world than in others. In particular, areas such as India and Africa, which were conquered and colonized by Europeans, were more deeply affected than areas like Persia or China, over which European influence was exercised indirectly through indigenous rulers.[13] Thus we shall award to each region a share of our attention proportional to that which it received from nineteenth-century imperialists.

This book makes no claim, then, to destroy other interpretations of nineteenth-century imperialism. Rather it aims to open new vistas and to provoke fresh thinking on this subject, by adding the technological dimension to the list of factors other historians have already explored.

## NOTES

1. D. K. Fieldhouse, *Economics and Empire 1830–1914* (Ithaca, N.Y., 1973), p. 3.

2. F. J. C. Hearnshaw, *Sea-Power and Empire* (London, 1940), p. 179.

3. J. A. Hobson, *Imperialism: A Study* (London, 1902). The debate has grown to the point of generating its own anthologies and historiographies. See, for example, Harrison M. Wright, ed., *The "New Imperialism": Analysis of Late Nineteenth Century Expansion*, 2nd ed. (Lexington, Mass., 1976); George H. Nadel and Perry Curtis,

eds., *Imperialism and Colonialism* (New York, 1964); and Ralph Austen, ed., *Modern Imperialism: Western Overseas Expansion and its Aftermath, 1776–1965* (Lexington, Mass., 1969). More analytical studies of the debate will be found in E. M. Winslow, *The Pattern of Imperialism* (London, 1948); George Lichtheim, *Imperialism* (New York and Washington, D.C., 1971); Roger Owen and Bob Sutcliffe, *Studies in the Theory of Imperialism* (London, 1972); and Benjamin Cohen, *The Question of Imperialism: The Political Economy of Dominance and Dependence* (New York, 1973). A detailed bibliography of imperialism will be found in John P. Halstead and Serafino Porcari, *Modern European Imperialism: A Bibliography,* 2 vols. (Boston, 1974); on the historiography of imperialism, see 1:32–37.

4. Ronald Robinson and John Gallagher with Alice Denny, *Africa and the Victorians: The Climax of Imperialism* (Garden City, N.Y., 1968), pp. 17–19.

5. Henri Brunschwig, *French Colonialism 1871–1914: Myths and Realities,* trans. William Glanville Brown (London, 1966).

6. Fieldhouse, pp. 81 and 463.

7. Fieldhouse, pp. 460–61.

8. Hans-Ulrich Wehler, *Imperialismus* (Berlin and Cologne, 1970), p. 13.

9. Hans-Ulrich Wehler, "Industrial Growth and Early German Imperialism," in Owen and Sutcliffe, pp. 72–73.

10. Rondo Cameron, "Imperialism and Technology," in Melvin Kranzberg and Carroll W. Pursell, Jr., eds., *Technology in Western Civilization,* 2 vols. (New York, 1967), 1:693.

11. Philip Curtin, Steven Feierman, Leonard Thompson, and Jan Vansina, *African History* (Boston, 1978), p. 448. See also David Landes, "The Nature of Economic Imperialism," *The Journal of Economic History* 21(1961):511 for a similar approach; the author recognizes the importance of technological factors, but does not explain or elaborate. In addition to these examples taken from the general literature, the technological factor appears in specialized writings on particular aspects of imperialism; see, for example, Philip Curtin, " 'The White Man's Grave': Image and Reality, 1780–1850," *Journal of British Studies* 1(1961):94–110 and *The Image of Africa: British Ideas and Actions 1780–1850* (Madison, Wis., 1964); Michael Gelfand, *Rivers of Death in Africa* (London, 1964); a series of articles on firearms in *The Journal of African History* 12(1971) and 13(1972); Michael Crowder, ed., *West African Resistance: The Military Response to Colonial Occupation* (London, 1971); and Henri

Brunschwig, "Note sur les technocrates de l'impérialisme français en Afrique noire," in *Revue française d'histoire d'outre-mer* 54(1967): 171–87. Such specialized works, however, are rare and the authors make no attempt to generalize about the role of technology in imperialism.

12. See, for example, Samuel Eliot Morison, *The European Discovery of America: The Northern Voyages* (New York, 1971), ch. 5: "English Ships and Seamen 1490–1600"; Joseph R. Levenson, ed., *European Expansion and the Counter-Example of Asia, 1300–1600* (Englewood Cliffs, N.J., 1967), ch. 1: "Technology"; Eugene F. Rice, *The Foundations of Early Modern Europe, 1460–1559* (New York, 1970), ch. 1: "Science, Technology and Discovery"; J. M. Parry, *The Establishment of the European Hegemony: 1415–1715: Trade and Exploration in the Age of the Renaissance* (New York, 1961), ch. 1: "The Tools of the Explorers: (i) Charts (ii) Ships (iii) Guns"; and Carlo Cipolla, *Guns and Sails in the Early Phase of European Expansion 1400–1700* (London and New York, 1965).

13. As for Latin America, its history in the nineteenth century, that is, in its post-independence or neo-colonial phase, is entangled with that of American expansion and resembles that of Africa and Asia since World War Two. To do it justice would require another book.

# PART ONE

# STEAMBOATS AND QUININE, TOOLS OF PENETRATION

# CHAPTER ONE

# Secret Gunboats
# of the East India Company

We have the power in our hands, moral, physical, and mechanical; the first, based on the Bible; the second, upon the wonderful adaptation of the Anglo-Saxon race to all climates, situations, and circumstances . . . the third, bequeathed to us by the immortal Watt. By his invention every river is laid open to us, time and distance are shortened. If his spirit is allowed to witness the success of his invention here on earth, I can conceive no application of it that would receive his approbation more than seeing the mighty streams of the Mississippi and the Amazon, the Niger and the Nile, the Indus and the Ganges, stemmed by hundreds of steam-vessels, carrying the glad tidings of "peace and good will toward men" into the dark places of the earth which are now filled with cruelty.[1]

Macgregor Laird

The steamboat, with its power to travel speedily upriver as well as down, carried Europeans deep into Africa and Asia. Few inventions of the nineteenth century were as important in the history of imperialism.

The idea of propelling a boat by the power of steam is an old one, dating back to the seventeenth century, before there were any working steam engines. In the late eighteenth century, many inventors in France, Britain, and the United States experimented with steam-driven vessels. Most English-speaking historians date the beginning of the age of steamboats from Robert Fulton's *Clermont,* which navigated the Hudson between New York and Albany in 1807, a commercial as well as a technical success.[2] Americans took to the new device with instant enthusiasm, for theirs was a land of great roadless distances crossed by broad rivers lined with trees which steamers could use for fuel. Within a decade the Hudson, Delaware, Potomac, and Mississippi and their tributaries all carried regular steamboat service.

Europeans took less eagerly to the new technology; tradition lay thicker, there were good roads, and fuel was costly. Yet soon the new invention appeared in European waters. Henry Bell's *Comet* brought regular steamboat service to the Clyde River in 1811. By 1815 ten steamboats operated on the Clyde and several more on other British rivers. By 1820 hundreds of steamboats plied the rivers and lakes of Europe, and a few were venturing out into the Mediterranean, the Channel, and the Irish Sea. From that point on, the history of steam vessels took a new turn. Small river craft became so common that they were taken for granted, as attention turned to the newer railroads and ocean-going steamships.

If henceforth steamboats on European rivers deserve little mention, it does not follow that European steamboats had little impact elsewhere. The revolution they wrought occurred in Asia and Africa, into whose interior they carried the power that European ships had possessed on the high seas for centuries. Indeed no single piece of equipment is so closely associated with the idea of imperialism as is the armed shallow-draft steamer, in other words, the gunboat.

The date of the gunboat's introduction is still a matter of debate. Antony Preston and John Major, historians of the gunboat era, date its birth to the Crimean War (1853–56) when the Royal Navy, suddenly forced to fight in the shallow waters of the Black and Baltic seas, ordered scores of such craft. Before that, they declare, "the word 'gunboat' was a generic term for a few minor vessels," as though only Royal Navy gunboats mattered.[3]

The naval historian Gerald Graham gives a more plausible date for the opening of the gunboat era. Speaking of Sino-British relations in the early nineteenth century, he notes:

> . . . the Royal Navy did not count as an instrument of diplomacy, because Peking had no notion of its true capability. . . . There was little or no hope in the 1820s that the Emperor and his Court could be made to see reason, and come to terms. If a final settlement were to be reached, it became gradually clear to Foreign Office officials that Peking would have to learn, through 'imposing demonstrations of force', the immense and devastating power that lay in the hands of the Mistress of the Seas. . . .
>
> Yet it was not until the beginning of the 1840s that the process of demonstrating power, of 'showing the flag' beyond coastal boundaries, became a practicable operation. The coming of steam, by making possible the penetration of Chinese rivers, opened up the interior to British warships. By 1842, steamers like the *Nemesis* could guide or pull battleships as far as Nanking on the Yangtse-kiang, more than 200 miles from the sea.[4]

The Opium War was the first event whose outcome was determined by specially built gunboats. Not by accident did gunboats appear in China at the right moment. They were the end product of a complex process of creation resulting from the confluence, in the mid-1830s, of two historical forces: the campaign to speed up communications between India and Britain by the use of steamers, and the innovative spirit of three individuals—Thomas Love Peacock and John and Macgregor Laird.

The first steamer in India was a small pleasure boat built in 1819 for the Nawab of Oudh. The Britons of Calcutta saw more practical applications for this new device, and in 1823 the Kidderpore dockyard built a 132-ton sidewheeler propelled by two 16-horsepower Maudslay engines. This vessel, the *Diana*, was put to work as a tugboat in Calcutta harbor, where sailing ships had great difficulty maneuvering. A year later she was joined by the *Pluto*, built in 1822 as a steam-dredge but then converted to a paddle-steamer.

These two boats were commercial failures but stimulated a tremendous interest in steam navigation. The British community of Calcutta, which felt isolated from its homeland—a round trip took a year or more—saw in steam a means of accelerating communications. In 1825 it offered a prize to the first steamer to make the trip in seventy days or less. In response, a group of steam enthusiasts in London built the *Enterprize*, a large ship with a weak and fuel-hungry engine. Although the *Enterprize* was deemed a failure, taking 113 days to make the trip to Calcutta, the voyage had several important consequences. By demonstrating that steamship technology was still too primitive to make ocean voyages commercially viable, or even worthy of government subsidy, the voyage shifted attention away from the unattainable "Cape Route" around Africa toward more immediate applications of steam: warfare, the navigation of the Ganges, and the "Overland Routes" to Britain via the Middle East.

In December 1824 the British East India Company and the kingdom of Burma had gone to war. Burma was separated from Bengal by almost impassable mountains, and the only easy access to the kingdom lay up the valley of the Irrawaddy River from Rangoon to the capital city of Ava. The war lasted two years and inflicted heavy casualties on both sides. The Burmese, despite their antiquated weapons, showed their skill as guerrillas in the swamps of the Irrawaddy. The British lost three quarters of their troops, most of whom were Indian sepoys, largely from disease.

The *Enterprize* ferried troops and mail from Calcutta to Rangoon. The *Pluto*, hastily armed with four brass 24-

pounders, participated in the attack on the Arakan peninsula. But the star of the war was the *Diana*. It was Captain Frederick Marryat, the naval commander and a well-known novelist, who had insisted that she be brought to Burma. On the Irrawaddy she towed sailing ships into position, transported troops, reconnoitered advance positions, and bombarded Burmese fortifications with her swivel guns and Congreve rockets.[5] The most important contribution of the *Diana* was to capture the Burmese *praus,* or war boats. One British officer wrote:

> The Burmese War boats were very fine in general about 90 feet long pulling 60 or 70 oars. Chiefs of consequence had gilt ones. They had heavy guns 6 or 9 pounders lashed in the Bows which frequently they could neither train or elevate. Their number of Boats were said to be very great but we never saw them in greater numbers than 25 or 30 at a time, and altogether they were a contemptible force and never evinced any spirit. They were extremely swift, far beyond any of our Boats, the Steam Vessel upon two or three occasions caught them by tiring their crews out. About 60 were captured during the war.[6]

As another observer pointed out, the race between the *Diana* and the *praus* was too unequal, for "the muscles and sinews of man could not hold out against the perseverance of the boiling kettle. . . ."[7] By February 1826 the *Diana,* which the Burmese called "fire devil," had pushed with the British fleet up to Amarapura, over 400 miles upriver. The king of Burma, seeing his capital threatened, sued for peace. Thus the East India Company secured Assam and acquired the Burmese provinces of Arakan and Tenasserim.[8]

It was quite by chance that the *Diana* took part in the Anglo-Burmese War. Yet, for all her heroics she only hastened the British victory, neither triggering the war nor determining its outcome. Her main contribution was to stimulate further interest in river steamers among the British in India. They no longer regarded steamers as just another type of vessel, but as an entirely new technology destined to enhance their power.

One application of steam power was internal. Transportation in British India was erratic and costly, for it depended on human beings, horses, and bullocks. Only the Ganges River, a natural highway through northern India that had been used by slow country boats from time immemorial, provided dependable, inexpensive transportation. Consequently, progress-minded Anglo-Indians greeted the advent of steamboats with the same enthusiasm exhibited by Americans opening up the Mississippi. So clearly was transportation in the interests of the Bengali government that steam service on the Ganges was from the beginning a political enterprise, a means of consolidating the British presence in Hindustan. In 1828, Governor-General Sir William Bentinck encouraged Captain Thomas Prinsep to begin surveying the Ganges River. Six years later a regular steam service between Calcutta and Allahabad was inaugurated, relying on the iron steamer *Lord William Bentinck,* a 120-foot sternwheeler with a speed of six to seven miles per hour. Within two years several other steamers cruised up and down the Ganges—the *Thames,* the *Jumna,* and the *Megna*—all pulling accommodation boats for passengers and barges for freight.[9]

These steamers were much faster than any country boat, taking only twenty days to travel from Calcutta to Allahabad in the rainy season, and twenty-four in the dry season. Yet they remained outside the economic life of the country because of their cost. A cabin from Calcutta to Allahabad cost 400 rupees (thirty pounds sterling), the same as a cabin on a steamer across the Atlantic or half the cost of a voyage from England to India. Deck passengers paid half as much, which was still a small fortune. As a result only government officials, bishops, planters, and Indian princes could afford the trip; in all of 1837 only 375 passengers traveled on a Ganges steamer.

Freight rates, equally outrageous at six to twenty pounds sterling per ton, limited the cargo on Ganges steamers to mostly government merchandise. Guns, flints, medicines, legal and official stationery, and delicate instruments were shipped upstream; in exchange, documents and the receipts from tax collectors flowed down to Calcutta. As for private freight, only the household goods of traveling officials and precious cargoes of silk, indigo, shellac, and opium could bear the cost of steam transport.[10]

The East India Company continued to operate steamers until 1844, after which several private companies joined the venture. In time, though, the Ganges became more difficult to navigate, because of massive deforestation, erosion and silting, and the diversion of water into irrigation canals. The other rivers of India, were too shallow or fickle to become major highways of steamer traffic. One ingenious engineer proposed in 1849 to build steamers with a twelve-inch draft and auxiliary wheels to overcome the sandbanks that blocked the Indus, the Nerbudda, and other rivers.[11] By then, however, a more efficient wheeled vehicle—the railroad—had eclipsed the steamboat.

While the *Diana* and the Ganges steamers were proving the practicability of steam navigation on rivers, the dream of rapid steamship service between India and Britain continued to tantalize the Anglo-Indians. In their minds, the failure of the *Enterprize* was only a temporary setback. As we shall see, the government of the Bombay Presidency launched the steamer *Hugh Lindsay,* which steamed to Suez and back in 1830. The Court of Directors of the East India Company, the Foreign Office, and the Admiralty did what they could to frustrate these attempts by Anglo-Indians to open steam communication independently. The steam lobby in India and its agents in London thereupon flooded Parliament with petitions and the newspapers with letters. In June 1834, in response to this campaign, the House of Commons appointed a Select Committee on Steam Navigation to India. Though concerned primarily with communication, this committee led directly to the beginning of the gunboat era.

The Select Committee of 1834 heard testimony from many witnesses, but three dominated the proceedings: Thomas Love Peacock, Francis Rawdon Chesney, and Macgregor Laird.

Peacock is well known to students of nineteenth-century English literature as a novelist and satirist, a close friend of Percy Bysshe Shelley, and the father-in-law of George Meredith. But he deserves equally to be known to maritime and technological historians. If he is not, it is because technological innovation often takes place in the discreet confines of workshops and offices, far from the public notice that attends literary and political lives.

In 1819, when Peacock first entered the East India Company, it had just created three new assistantships in the office of the Examiner of Correspondence: justice, revenue, and public works. After a two-year apprenticeship and some difficult examinations, Peacock became assistant to the examiner, in charge of public works.[12]

In 1828, Thomas Waghorn, an officer who had been favorably impressed with the *Diana* while serving in the Burma War, arrived in London on behalf of the merchants of Calcutta and Madras, hoping to persuade the East India Company to establish regular steam communication with India. John Loch, chairman of the Court of Directors, asked Peacock to look into the matter.[13] In September 1829, Peacock drew up a "Memorandum respecting the Application of Steam Navigation to the internal and external Communications of India." In it he argued that steamers should be used between Britain and India. Of the three possible routes—around Africa, via the Red Sea, and via Syria and the Euphrates—he preferred the Euphrates one, because river steamers were more reliable than seagoing steamships, the Euphrates had been navigated successfully since antiquity, and, most important, if Britain did not secure Mesopotamia, Russia would:

> The Russians have unlimited resources in coal, wood, iron, cattle and corn. They have now steamboats on the Volga and the Caspian Sea. They will have them before long on the Sea of

24

Aral and the Oxus, and in all probability on the Euphrates and the Tigris. It is not our navigating the Euphrates that will set them the example. They will do every thing in Asia that is worth the doing, and that we leave undone.[14]

In 1833, Governor-General Bentinck had sent Lieutenant Alexander Burnes to London to bring the question of steamers before Parliament. On December 12, 1833, Burnes wrote Bentinck:

I have not failed in all my communications with the authorities in England to inform them of your Lordship's extreme anxiety regarding a more extensive use of steam in India and a permanent communication by means of it with Europe. Mr. Grant was reading the minute last time I saw him. The chairman of the court appears fully sensible of the great advantages of it and declares his utmost readiness to comply with the suggestions if the finances of India permit of it. I was sorry to hear a gentleman at the India House, and a man of some consequence there, Mr. Peacock, expressing his opinions of the disadvantages of a more rapid communication with India. He thought that the *brief* and garbled accounts which the house would receive in hurried letters would be most unsatisfactory, but this is the voice of one man and I cannot doubt that which is so apparently and obviously necessary will no longer be denied.[15]

Perhaps Peacock truly was concerned, in his capacity as assistant examiner of correspondence, with the quality of writing in the letters he would have to read. More likely the satirist was just teasing the poor lieutenant.

Before the Select Committee met, Peacock made the acquaintance of Artillery Captain Francis Rawdon Chesney, who had recently returned from a tour of the Middle East. Back in 1829, when Peacock was writing his "Memorandum," Sir Robert Gordon, British ambassador in Constantinople, had sent Chesney to Egypt on a political mission. There, Consul-General John Barker handed Chesney a series of questions drawn up by Peacock as to the relative merits of the Egyptian and Mesopotamian routes with respect to time of travel, security, trade, navigation, fuel supplies, and local inhabitants.

The following year Chesney set out alone to explore the Euphrates. When he returned to England, King William IV told him of

> serious apprehensions caused by the presence of the Russian fleet at Constantinople, as well as by the gradual advance of that Power toward the Indus, and the consequent necessity of strengthening Persia; adding, that as an additional security to our position it might be advisable to carry out [Chesney's] suggestion by adding a steam flotilla to the Bombay Marine.[16]

Peacock encouraged Chesney to write a report stressing the advantages of the Euphrates route. Foreign Secretary Viscount Palmerston, who was pro-Ottoman, anti-Russian, and anti-Egyptian, received the report favorably.[17]

In March 1834, Chesney and Peacock held several conversations in which, according to Chesney, they agreed that a Suez Canal would be "quite practicable." They also discussed the merits of high- versus low-pressure engines, and of towing a second steamer as opposed to an accommodation boat. Chesney came away impressed that the East India Company's directors were more enterprising than the government, and "more than half inclined to go to the expense of opening the Euphrates themselves."[18]

The third major witness, Macgregor Laird, was both a noted explorer and an expert on steamboats. His father, William Laird, had moved from Greenock, Scotland, to Liverpool in 1822, when Macgregor was thirteen years old. In 1824, William founded the Birkenhead Iron Works, which he renamed William Laird and Son in 1828 when Macgregor's older brother John became a partner. The following year the Lairds built their first boat, the lighter *Wye,* for the Irish Inland Steam Navigation Co. Then in 1830 came an order from the City of Dublin Steam Packet Co. for a 133-foot, 184-ton iron paddle-steamer named *Lady Lansdowne,* for use on the Shannon River. The Lairds' third boat, the *John Randolph,* was sent in sections to Savannah, Georgia; she was the first iron boat in America.

During those years, Britain's iron production was fast increasing, thanks in part to Nielson's hot-blast process. Meanwhile, British forests had become inadequate to meet the needs of the wooden shipbuilding industry, and foreign timber supplies were judged too vulnerable. It was a propitious time for an iron works to enter the shipbuilding business, except for one drawback; the public and the major purchasers of ships remained skeptical about the seaworthiness of both steamers and iron ships. It would take spectacular examples to dispel their doubts.

In 1832, Macgregor Laird organized an expedition to Africa in two small steamers and a sailing ship. One of his three ships, the *Alburkah,* was made of iron. When she sailed from Liverpool she aroused comment and ridicule, for never before had an iron steamer ventured out into the ocean. Laird wrote:

> It was gravely asserted that the working in a sea-way would shake the rivets out of the iron of which she was composed: the heat of the tropical sun would bake alive her unhappy crew as if they were in an oven; and the first tornado she might encounter would hurl its lightning upon a conductor evidently set forth to brave its power.[19]

Yet the *Alburkah* performed very well and never even leaked, despite repeated groundings.

While Macgregor Laird was away in Africa, his father and brother built an iron steamer, the *Garryowen,* for the City of Dublin Steam Packet Co. In 1834, during a test cruise, she was grounded in a storm that would have wrecked a ship of wood. The survival of the *Alburkah* and the *Garryowen* did more than any theory to gain the public's confidence in iron ships.[20]

Macgregor Laird returned to England in early 1834, ill and financially ruined, but famous. It was in those months before the Select Committee met that he made friends with Thomas Love Peacock.[21]

Peacock's testimony fills half the "Report of the Select Committee on Steam Navigation to India." He recommended the Overland Routes as preferable to the Cape Route, but of the two Overland Routes, he preferred the Mesopotamian Route as being less expensive and more easily navigated than the Red Sea. He offered a wealth of details on the navigation of the Indian Ocean, the Red Sea, and the Euphrates, much of it based on ancient sources and travelers' reports. He waxed most eloquent, however, on the need to prevent a Russian invasion of Mesopotamia:

> The first thing the Russians do when they get possession of, or connexion with, any country, is to exclude all other nations from navigating its waters. I think therefore it is of great importance that we should get prior possession of this river.

When asked: "Is it your opinion that the establishment of steam along the Euphrates would serve in any respect to counteract Russia?" Peacock replied: "I think so, by giving us a vested interest and a right to interfere."[22]

When in the course of his testimony, Peacock was asked technical questions about steam navigation, he deferred to the expertise of Macgregor Laird. Thus on the subject of the proper ratio of power to tonnage in steamers, he declared:

> Mr. M'Gregor Laird, who has had as much experience as any man in this country, thinks he would not have more than two and a half for the proportion of tonnage to measurement.[23]

The committee began by questioning Laird on his Niger Expedition—his ships, their qualities at sea and on the river, the diseases his men suffered. One subject they were especially interested in, and on which he was most eager to speak, was the relative merits of his two Niger steamers, the *Alburkah* and the *Quorra,* and of iron and wooden ships in general. In his estimation the *Alburkah* had all the virtues: "She was a much livelier vessel, she did not labour so much. . . . I think myself that the principal reason of an iron vessel being so much healthier is on account of her coolness and her freedom from all manner of smell. . . . It is impossible for the iron vessel to be struck by lightning. . . ." He went on to describe many other advantages of iron: freedom from vermin, decay, corrosion, and leakage; lightness, more cargo space, greater speed with less fuel; resistance to puncture when grounding, to breaking on rocks, to flexing in heavy seas, to splintering from the impact of cannonballs; and the possibility of building an iron ship with watertight bulkheads, making it much safer than a ship of wood. As for the advantages of wood, he knew of none.

He was then asked about building an iron steamer. He went into great detail on the proper ratios of the outer dimensions, the power of the engine, cylinder diameter, piston stroke, type, consumption and supply of coal, copper versus iron boilers, steam pressures, oscillating cylinders, variable paddles, speed, accommodations, cargo capacity, range, and much else. For river steamers he recommended that the keel be flat in the midships section, with parabolic sections fore and aft; that the paddle shaft be placed one third the way from the stern at the point of greatest beam; and that a single engine be placed above deck. He described the ideal river steamer as being 110 feet long by 22 feet wide by 7½ feet high with a 3-foot draft, powered by a 50-horsepower engine. On the subject of steam pressure he was a conservative, like most British engineers of his day, and recommended a limit of 10 pounds per square inch. "Are not the American vessels working as high as 30 pounds an inch?" he was asked. "Yes," he replied, "but they kill about a thousand people every year."

The questioning then turned to the specifics of the Mesopotamian Route. He gave estimates of the cost of steamers of various sizes and of providing regular service via the Euphrates and via the Red Sea. He estimated that steamers for the Euphrates could be built in six months. He advised sending them in sections to Basra on the Persian Gulf or to Bombay, and assembling them there and steaming up rather than down the river. His testimony was that of an interested party, yet from the questions he was asked it is clear that his expertise was taken very seriously.[24]

Chesney's testimony dealt mainly with the navigability of the Euphrates and with the geography and ethnography of the area. On one point only did his ideas differ significantly from Laird's; he advised sending the steamers in sections overland to a point on the upper Euphrates, and steaming downriver rather than up. On that point, his advice prevailed; it turned out later to be a mistake. Otherwise the testimony of the three key witnesses dovetailed nicely. To those familiar pressures for imperial expansion, the clamor from the periphery of empire and the lobbying of interested business groups, Peacock added a political justification for sending steamers to the Euphrates: the specter of Russia, that axiom of Palmerstonian foreign policy. Behind these drives, however, lay a new opportunity opened up by technological innovation; this was Laird's contribution. The Middle East, which empires of the age of sail had found to be only an obstacle, had now become worth coveting.

The Select Committee of 1834 had just the effect that the Anglo-Indian steam enthusiasts had hoped for. It gave the approval and financial backing of the British government to both Overland Routes—the Egyptian and the Mesopotamian. The Select Committee recommended that Parliament vote the sum of £20,000 to send two steamers to the Euphrates. Captain Chesney, whose knowledge of Middle Eastern geography had impressed the members, was given command of the expedition. From the Lairds the East India Company ordered two boats resembling the description Macgregor Laird had given of the ideal river steamer. The *Euphrates* was 105 feet long, the *Tigris* 87; both were made of iron, drew less than 3 feet of water, and carried several small artillery pieces. Their crews were trained at the Laird shipyard.[25] They were, in the words of Captain Chesney, "the first of the flat armed steamers, whose services have been so important in the rivers of Asia."[26]

The Euphrates Expedition took far longer than expected. The steamers were taken in sections by sailing ship to the Bay of Antioch. It then took over a year—from the end of 1834 to April 1836—to cart the pieces to Bir on the Euphrates and re-assemble them. Mishaps attended the expedition, including harassment by Egyptian saboteurs and the loss of the *Tigris* in a storm. Steaming down the river to the Persian Gulf took another year, for Chesney was more interested in surveying the river and its inhabitants in minute detail than in setting speed records. In mid-1837 he returned to England.[27]

The Euphrates Expedition failed to achieve its purpose; by the time the members returned, seagoing steamships had replaced river steamers as the preferred means of communication to India. Unexpected consequences, however, flow from such failures: the Select Committee of 1834, by bringing together Peacock and the Lairds, had launched the gunboat era.

In 1836, upon the death of James Mill, Peacock was promoted to chief examiner of correspondence, one of the highest positions in the East India Company. Furthermore, he had the support and friendship of the powerful John Cam Hobhouse, president of the Board of Control, the body through which Parliament supervised the affairs of India.[28] The responsibility for steam navigation was now entirely his. In later years John Laird modestly attributed to Peacock the credit for developing the gunboat, although he and his brother also deserve a share:

> . . . the late Mr Peacock was instrumental in extending and improving Steam Navigation at a time when long voyages were considered impracticable, and also in taking the responsibility of adopting the plans suggested to him for constructing a new Class of Iron men of war of light draught, but of sufficient strength to

enable them to carry guns of very heavy Calibre, a system very generally adopted since by most nations.[29]

The East India Company now became Laird's most important customer. Of the twenty-seven iron steamers built at the Birkenhead Iron Works before 1841, twelve were for the company.[30]

In 1837 the company bought a Laird steamboat originally built for an American customer. This boat, the *Indus,* was 115 feet long by 24 feet wide, and had a 60-horsepower engine. She was shipped in sections to Bombay, reassembled, and sent out to patrol the Indus River.[31] She was followed a year later by two more steamboats, the 132-foot *Comet* and the 102-foot *Meteor.* They drew too much water for the shallow river, however, and their engines were barely able to push them against the swift currents.[32]

In Mesopotamia, meanwhile, Chesney had handed over the command of the *Euphrates* to Lieutenant Henry Lynch, who proceeded to survey the Tigris River and show the British flag as far as Baghdad.[33] In 1838 another Russian alarm arose, coupled with a growing threat from the French-backed Mehemet Ali of Egypt. The East India Company now undertook to strengthen its position in the area. In August 1838, Peacock wrote a "Memo on the mode of obtaining small Steamers for the Euphrates & Tigris Rivers":

> The considerations also on which the policy of employing additional Steamers could be recommended, are themselves of a nature, which it would not be desirable to disclose while they are in progress; when the object to be attained is to establish an influence in that quarter to the exclusion of that of other European Powers.
>
> In this point of view it would be very desirable if it be practicable to obtain the Steamers by an order from the Secret Committee, instead of from the Court.[34]

This Secret Committee was composed of three directors whose duty it was to transmit the orders of the Board of Control, in particular of its president, John Cam Hobhouse, to the governor-general of India, bypassing the other directors.[35]

In September there followed a letter from Hobhouse, recommending that the Secret Committee purchase two or three small steamers for the Tigris and the Euphrates.[36] On October 5, the Secret Committee agreed to place an order with Laird the following February and March for three armed iron steamers powered by Forrester engines.[37] The following May, while the steamers were being built, the Secret Committee decided that they should ostensibly be sent to South America, but with secret instructions.[38] Behind this veil of secrecy, the *Nimrod, Nitocris,* and *Assyria* were shipped in sections to Basra on the Persian Gulf. John Laird later wrote:

> These Vessels were 100 feet long, 18 feet beam, & 40 Horse Power, and were shipped in pieces, and workmen sent with them to put them together on arrival at the mouth of the Euphrates. Engineers, Carpenters, Joiners, Boiler-Makers, and Iron Shipbuilders were sent out, and the Boats were at work on the River before it was generally known such an expedition had left England.[39]

They joined the *Euphrates* in patrolling the rivers of Mesopotamia. They carried no mail or passengers. Rather, their purpose was to demonstrate Britain's support for Ottoman rule and to keep the Russians at bay.

By December 1838 the company had therefore ordered or sent out eight iron gunboats—five for the Tigris and Euphrates and three for the Indus. Furthermore, the Indian Navy, now reinforced with steamships, commanded the Indian Ocean, the Red Sea, and the Persian Gulf. Wherever along the southern rim of Eurasia Britain's ships and boats could reach, she had little to fear from any other power.

Yet at that moment the Secret Committee of the East India Company decided to order five more gunboats, and in the following month a sixth. This new series of steamers was to be of the same general dimensions as the previous boats—100 to

130 feet long, 18 to 26 feet wide but with two important innovations: sliding keels that could be adjusted for deep or
shallow water and engines of 70 to 110 horsepower, compared
to 20 to 60 for previous gunboats in the area.[40]

In the minutes of the Secret Committee, these new boats
were described as destined for the Indus River. However,
there is good reason to doubt the sincerity of this intention.
On April 1, 1839, George Eden, earl of Auckland, governor-
general of India, wrote to Hobhouse:

> The Court is most liberal on the subject of Steamers and when
> the keel which has been announced shall be completed we shall
> be strong indeed. I would suggest that two of the vessels which
> will be adapted as well for sea as for river navigation should be
> sent to Calcutta. They would be invaluable should that reckless
> savage of Ava [the King of Burma] force us into hostilities with
> him. And I suspect that their draft of water may be rather too
> great for the Bar of the Indus.[41]

And on June 14 he wrote Hobhouse again:

> I am happy to learn from Bombay that they are putting together
> for the Indus two of the iron steamers which you have sent out
> and I shall be satisfied with these for the present and would
> have any others sent to Bengal, where they will be of infinite
> and certain use while the extensive application of steam to the
> navigation of the Indus must for a time be speculative.[42]

Perhaps, as Auckland implied, the company had been over-
enthusiastic in ordering six more steamers for the Indus when
three sufficed. More likely, Peacock and the Secret Committee
used the Indus as a cover for another purpose that they did
not wish to reveal.

Two of the gunboats, the *Pluto* and the *Proserpine,* were ordered from Ditchman and Mare at Deptford on the Thames.
They were small—less than 200 tons—and made of wood, with
sliding keels. Peacock himself supervised their construction.
The other four were to be built of iron by John Laird at Birkenhead. The *Ariadne* and the *Medusa* were of medium size—

432 tons, 139 feet long by 26 feet wide inside the paddle-boxes—with 70-horsepower engines and two 24-pounder cannons apiece. The *Phlegethon* was larger: 510 tons, 161 feet long by 26 feet wide, with a 90-horsepower engine.

The one that aroused the most interest, though, was the *Nemesis*. The largest iron ship ever built thus far, she was also the first of the new series to be launched. The appearance in January 1840 of a heavily armed, radically new kind of craft privately registered in the name of John Laird aroused a flurry of speculation.

When the *Nemesis* paid a visit to Greenock, the local *Shipping Gazette* wrote: "She will, it is said, clear out for Brazil; but her ultimate destination is conjectured to be the eastern and Chinese seas."[43] On February 27, Hobhouse wrote Palmerston:

> An Armed iron vessel, called the Nemesis, which has been provided by the Secret Committee of the East India Company under the Authority of the Commissioners for the affairs of India, for the service of the Government of India, is about to proceed to Calcutta. . . .
> It is desirable that the destination of the Nemesis, and the authority to which she belongs, should not be mentioned.[44]

A similar request went out from the Secret Committee to the Admiralty.[45] On March 28, wrote her captain William Hall, "she was cleared out for the Russian port of Odessa, much to the astonishment of every one; but those who gave themselves time to reflect, hardly believed it possible that such could be her real destination."[46] Two days later *The Times* commented:

> Sailed to-day the Nemesis, private armed steamer, Hall master, destination unknown. It is said this vessel is provided with an Admiralty letter of license, or letter of marque; if so, it can only be against the Chinese; and for the purpose of smuggling opium she is admirably adapted. Others conjecture she is going to Circassia for sale, as she is also well suited to defend a port, or for offensive operations in shallow waters.[47]

Once underway, Captain Hall told his crew that they were heading for Malacca via Capetown and Ceylon. Only when he reached Ceylon in October did he receive the order to proceed to his real destination: Canton.[48]

The secrecy that surrounded the new gunboats reveals some of the contradictions between technological innovation and military power in the mid-nineteenth century. Military power, to Britain at that time, meant the Royal Navy. Of all the institutions in Britain, the Navy was among the most conservative in technological matters. It long refused to purchase steamers, arguing that their wheels were too vulnerable to enemy fire, took up too much space needed for cannons, and dragged when under sail. Worse, the engines' voracious appetite for coal limited their range and made them too costly for the ever-so-parsimonious lords of the Admiralty. As for other Navy men, they had, in Gerald Graham's words, "no love for foul-smoking steamers and dirtier men."[49]

Iron was even more anathema to the Navy. Not until 1845 did it purchase its first iron ship. Even then Auckland, then first lord of the Admiralty, could write:

> . . . iron is a material which may very advantageously be used for the construction of vessels intended to act in shallow waters, and indeed for vessels intended to be used for many other purposes, but I apprehend that a construction of this material is not adapted for the general purposes of war . . . an iron vessel could ill stand a heavy broadside.[50]

As late as 1851 the Admiralty informed the Peninsular and Oriental steamship line that it would not approve any ship "if built of iron or of any material offering so ineffectual a resistance to the striking of shot."[51]

On the high seas, as long as no enemy challenged Britain's naval supremacy, such a policy was no doubt quite sensible. It was most ill-suited, however, to the shallow waters of imperialist warfare.[52] For Peacock and the Lairds to have presented their machines as men-of-war would have risked an adverse reaction from those in power, who had grown up with a deep reverence for wooden sailing ships of the line. This then is the reason for the secrecy that surrounds the gunboats sent to China. They were not camouflaged behind false destinations— the Indus, Brazil, Calcutta, Odessa—to fool the Chinese; they were hidden from the Admiralty and the government. Before the age of military research and development, technological innovation often had to sneak through the back door.

## NOTES

1. Macgregor Laird and R. A. K. Oldfield, *Narrative of an Expedition into the Interior of Africa, by the River Niger, in the Steam-Vessels Quorra and Alburkah, in 1832, 1833, and 1834*, 2 vols. (London, 1837), 2:397–98.
2. The early history of steamboats has been told many times, though the accounts lack consistency. See in particular W. A. Baker, *From Paddle-Steamer to Nuclear Ship: A History of the Engine-Powered Vessel* (London, 1965), pp. 10–12; Ambroise Victor Charles Colin, *La navigation commerciale au XIXe siècle* (Paris, 1901), pp. 37–38; Maurice Daumas, ed., *Histoire générale des techniques*, 3 vols. (Paris, 1968), 3:332–33; Eugene S. Ferguson, "Steam Transportation," in Melvin Kranzberg and Carroll W. Pursell, Jr., eds., *Technology in Western Civilization*, 2 vols. (New York, 1967), 1:284–91; Duncan Haws, *Ships and the Sea: A Chronological Review* (New York, 1975), pp. 100–01; F. J. C. Hearnshaw, *Sea-Power and Empire* (London, 1940), pp. 190–91; George W. Hilton, Russell A. Plummer, and Joseph Jobé, *The Illustrated History of Paddle Steamers* (Lausanne, 1977), pp. 9–12; George Gibbard Jackson, *The Ship Under Steam* (New York, 1928), ch. 1; Thomas Main (M.E.), *The Progress of Marine Engineering from the Time of Watt until the Present Day* (New York, 1893), pp. 10–16; Michel Mollat, ed., *Les origines de la navigation à vapeur* (Paris, 1970); George Henry Preble, *A Chronological History of the Origin and Development of Steam Navigation*, 2nd ed. (Philadelphia, 1895), pp. 119–25; Hereward Philip Spratt, *The Birth of the Steamboat* (London, 1958), pp. 17–40; Joannès Tramond and André Reussner, *Eléments d'histoire maritime et coloniale*

(*1815–1914*) (Paris, 1924), p. 50; David B. Tyler, *Steam Conquers the Atlantic* (New York, 1939), p. 112; and René Augustin Verneaux, *L'industrie des transports maritimes au XIXe siècle et au commencement du XXe siècle*, 2 vols. (Paris, 1903), 2:16–23.

3. Antony Preston and John Major, *Send a Gunboat! A Study of the Gunboat and its Role in British History, 1854–1904* (London, 1967), pp. 3 and 191 ff.

4. Gerald S. Graham, *The China Station: War and Diplomacy 1830–1860* (Oxford, 1978), pp. 18–19.

5. Congreve rockets are best known to Americans for the verses they inspired: "And the rockets' red glare, the bombs bursting in air/ Gave proof through the night that our flag was still there."

6. "Captain Chad's Remarks on Rangoon and the War in Ava 1824–1825–1826" (written June 3, 1827), in British Library Add. MSS 36,470 (Broughton Papers), pp. 100–101.

7. *United Service Journal* 2(1841):215, quoted in Gerald Graham, *Great Britain in the Indian Ocean: A Study of Maritime Enterprise, 1810–1850* (Oxford, 1968), p. 352.

8. On the early steamers of India and their use in the Anglo-Burmese War, see Graham, *Indian Ocean*, pp. 345–58; Henry T. Bernstein, *Steamboats on the Ganges: An Exploration in the History of India's Modernization through Science and Technology* (Bombay, 1960), pp. 28–31; H. A. Gibson-Hill, "The Steamers Employed in Asian Waters, 1819–39," *The Journal of the Royal Asiatic Society, Malayan Branch*, 27 pt. 1 (May 1954): 127–61; D. G. E. Hall, *Europe and Burma: A Study of European Relations with Burma to the Annexation of Thibaw's Kingdom, 1886* (London, 1945), p. 115; Col. W. F. B. Laurie, *Our Burmese Wars and Relations with Burma: Being an Abstract of Military and Political Operations, 1824–25–26, and 1852–53* (London, 1880), pp. 71–72 and 119–20; John Fincham, *History of Naval Architecture* (London, 1851), pp. 294–96; Charles Rathbone Low, *History of the Indian Navy (1613–1863)*, 2 vols. (London, 1877), 1:412n; *The Mariner's Mirror: The Journal of the Society for Nautical Research*, 30(1943):223 and 31(1944):47; and Oliver Warner, *Captain Marryat, A Rediscovery* (London, 1953), p. 67. There is also some information on the subject in Preble, pp. 76–77 and 120, but it is unreliable.

9. The definitive work on this subject is Henry T. Bernstein's *Steamboats on the Ganges*. See also Gibson-Hill, pp. 121–23; A. J. Bolton, *Progress of Inland Steam Navigation in North-East India from 1832* (London, 1890), p. 330; and J. Johnston, *Inland Navigation on the Gangetic Rivers* (Calcutta, 1947), p. 28.

10. Bernstein, ch. 5; Johnston, pp. 28–29.

11. John Bourne, C. E., *Indian River Navigation: A Report Addressed to the Committee of Gentlemen Formed for the Establishment of Improved Steam Navigation upon the Rivers of India, Illustrating the Practicality of Opening up Some Thousands of Miles of River Navigation in India, by the Use of a New Kind of Steam Vessel, Adapted to the Navigation of Shallow and Shifting Rivers* (London, 1849).

12. Edward Strachey, "Recollections of Peacock," in Thomas Love Peacock, *Calidore & Miscellanea*, ed. by Richard Garnett (London, 1891), p. 15; Carl Van Doren, *The Life of Thomas Love Peacock* (London and New York, 1911), pp. 212–14; Edith Nicolls, "A Biographical Notice of Thomas Love Peacock, by his Granddaughter," in Henry Cole, ed., *The Works of Thomas Love Peacock*, 3 vols. (London, 1875), 1:xxxvii; Sylva Norman, "Peacock in Leadenhall Street," in Donald H. Reiman, ed., *Shelley and His Circle*, 4 vols. (Cambridge, Mass., 1973), 4:709–23.

13. "Biographical Introduction," in Thomas Love Peacock, *Works* (The Halliford Edition), ed. by Herbert Francis Brett-Smith and C. E. Jones, 10 vols. (London and New York, 1924–1934), 1:clix–clx; Felix Felton, *Thomas Love Peacock* (London, 1973), pp. 230–31. See also Thomas Love Peacock, *Biographical Notes from 1785 to 1865* (London, 1874), and Richard Garnett, "Peacock, Thomas Love," in *Dictionary of National Biography*, 44:144–47.

14. The 1829 Memorandum can be found in *Parliamentary Papers* 1834 (478.) XIV, pp. 610–18.

15. Cyril Henry Philips, ed., *The Correspondence of Lord William Cavendish Bentinck, Governor-General of India, 1828–1834*, 2 vols. (Oxford, 1977), 2:1164–65.

16. Francis Rawdon Chesney, *Narrative of the Euphrates Expedition Carried on by Order of the British Government During the Years 1835, 1836, and 1837* (London, 1868), pp. 4 and 145–46.

17. Chesney, p. 143; Halford Lancaster Hoskins, *British Routes to India* (London, 1928), pp. 154–55.

18. Stanley Lane-Poole, ed., *The Life of the Late General F. R. Chesney Colonel Commandant Royal Artillery D.C.L., F.R.S., F.R.G.S., etc. by his Wife and Daughter*, 2nd ed. (London, 1893), pp. 269–70.

19. Laird and Oldfield, 1:7.

20. On the beginnings of the Laird firm, see Cammell Laird & Co. (Shipbuilders & Engineers) Ltd., *Builders of Great Ships* (Birkenhead, 1959), ch. 1; P. N. Davies, *The Trade Makers: Elder Dempster*

*in West Africa, 1852–1972* (London, n.d.), pp. 35–39 and 404 table 9;
Stanislas Charles Henri Laurent Dupuy de Lôme, *Mémoire sur la
construction des bâtiments en fer, adressé à M. le ministre de la
marine et des colonies* (Paris, 1844), pp. 4–6 and 117–19; Fincham,
pp. 386–87; Francis E. Hyde, *Liverpool and the Mersey: An Economic History of a Port 1700–1970* (Newton Abbot, 1971), pp. 52
and 84; "Laird, John," in *Dictionary of National Biography*, 11:
406–07; "Laird, Macgregor," in *Dictionary*, 11:407–08; and Tyler,
pp. 30–36, 112, and 169. In the India Office Records and Library are
various letters and reports concerning the Lairds; see L/MAR/C
vol. 583 pp. 217 and 223 and vol. 593 pp. 695–96.

21. Nicolls in Cole, 1:xxxviii. See also Arthur B. Young, *The Life
and Novels of Thomas Love Peacock* (Norwich, Eng., 1904), pp. 26–
27; and Diane Johnson, *The True History of the First Mrs. Meredith
and Other Lesser Lives* (New York, 1972), p. 60. Peacock and Laird
were to become relatives in 1844 when Peacock's daughter Mary
Ellen married a brother of Macgregor Laird's wife.

22. "Report from the Select Committee on Steam Navigation to
India, with the Minutes of Evidence, Appendix and Index," pp.
9–10, in *Parliamentary Papers* 1834 (478.) XIV, pp. 378–79. Peacock
also presented the committee with a series of papers relevant to the
issue of steam navigation: a short memorandum dated December
1833, quotations from travelers' accounts, and so on. There is a sum-
mary of the report in *Edinburgh Review* 60(Jan. 1835): 445–82,
probably written by Peacock himself.

23. Report, p. 6, in *Parliamentary Papers* 1834 (478.) XIV, p. 375.

24. Report, pp. 56–70 in *Parliamentary Papers* 1834 (478.) XIV,
pp. 425–39.

25. Chesney, *Narrative*, p. 154; Hoskins, p. 164; Gibson-Hill,
p. 123; Dupuy de Lôme, p. 6; and "List of Iron Steam and Sailing
Vessels Built and Building by John Laird, at the Birkenhead Iron
Works, Liverpool," in India Office Records, L/MAR/C 583 p. 217.
The *Euphrates* had a fifty-horsepower engine; there is much disagree-
ment over the power of the *Tigris*, with estimates ranging from
twenty to forty horsepower.

26. Francis Rawdon Chesney, *The Expedition for the Survey of
the Rivers Euphrates and Tigris, Carried on by Order of the British
Government, in the Years 1835, 1836, and 1837; Preceded by Geographical and Historical Notices of the Regions Situated Between
the Rivers Nile and Indus*, 2 vols. (London, 1850), 1:ix.

27. On the Euphrates Expedition, see also Ghulam Idris Khan,
"Attempt at Swift Communication between India and the West be-

fore 1830," *The Journal of the Asiatic Society of Pakistan* 16 no. 2 (Aug. 1971):124–27; John Marlowe, *World Ditch: The Making of the Suez Canal* (New York, 1964), p. 34; Low, 2:31–41; and Gibson-Hill, p. 123. The expedition also cost more than anticipated. In 1836, Parliament voted another £8,000 on the understanding that the company would also contribute £8,000 to complete the expedition; see *Parliamentary Papers* 1836 (159.) XXXVIII p. 418. Willson Beckles says that the expedition cost a total of £29,637 10s 3½d after deducting £10,360 12s 9d for steamers, arms, ammunition, instruments, and stores taken over by the company; see his *Ledger and Sword; or, The Honourable Company of Merchants of England Trading to the East Indies (1599–1874)*, 2 vols. (London, 1903), 2:393.

28. Hobhouse wrote of Peacock: "My intercourse with that most accomplished scholar, and most amiable man, has been one of the principle [sic] charms and resources of my declining years." Sir John Cam Hobhouse, First Baron Broughton, *Recollections of a Long Life, with Additional Extracts from his Private Diaries*, 6 vols. (London, 1910–11), 5:184. See also Young, p. 28.

29. John Laird, "Memorandum as to the part taken by the late Thomas Love Peacock Esq in promoting Steam Navigation" (1873), MS Peacockana 2 in The Carl H. Pforzheimer Library, New York. I am indebted to The Carl and Lily Pforzheimer Library, Inc. on behalf of the Carl H. Pforzheimer Library, and to Dr. Donald H. Reiman and Dr. Mihai H. Handrea, for permission to consult this document.

30. The financial side of these purchases is unclear. We know that much of the money came from the government under the rubric of estimates for steam communication with India, as follows:

| Year | Amount | in Parliamentary Papers |
|------|--------|-------------------------|
| 1834 | £20,000 | (492.) XLII. 459 |
| 1836 | 8,000 | (159.) XXXVIII. 418 |
| 1837 | 37,500 | (445.) XL. 401 |
| 1837–38 | 50,000 | (313.) XXXVII. 386 |
| 1839 | 50,000 | (142–IV) XXXI. 684 |
| 1840 | 50,000 | (179–IV) XXX. 859 |

Generally, the East India Company was expected to pay half the cost of steam communication east of Alexandria.

31. John Laird, "Memorandum."

32. Jean Fairley, *The Lion River: The Indus* (New York, 1975), pp. 222–25.

33. Lynch and his brother opened a trading house in Baghdad in 1840, bought the *Euphrates* from the East India Company, and later founded the "Euphrates and Tigris Steam Navigation Company"; see Zaki Saleh, *Britain and Mesopotamia (Iraq to 1914): A Study in British Foreign Affairs* (Baghdad, 1966), p. 179.

34. India Office Records, L/P&S/3/4 pp. 25–31.

35. See Robert E. Zegger, *John Cam Hobhouse: a Political Life 1819–1852* (Columbia, Mo., 1973), pp. 249–52.

36. India Office Records, L/P&S/3/4 pp. 33–38.

37. India Office Records, L/P&S/3/4, pp. 73–76.

38. India Office Records, L/P&S/3/4, pp. 215–18.

39. John Laird, "Memorandum." See also Dupuy de Lôme, pp. 4–5.

40. India Office Records, L/P&S/3/4, pp. 109–36: Minutes of the Secret Committee dated December 1838 and January 1839 concerning steamers for the Indus.

41. Broughton Papers, British Library Add. MSS 36,473, p. 446.

42. Broughton Papers, 36,474, p. 109.

43. *The Nautical Magazine and Naval Chronicle for 1840*, pp. 135–36.

44. India Office Records, L/P&S/3/6, p. 167.

45. India Office Records, L/P&S/3/6, pp. 619–29.

46. William H. Hall (Capt. R.N.) and William Dallas Bernard, *The Nemesis in China, Comprising a History of the Late War in that Country, with a Complete Account of the Colony of Hong Kong*, 3rd ed. (London, 1846), p. 6.

47. *The Times* (March 30, 1840), p. 7.

48. Hall and Bernard, pp. 18 and 61; India Office Records, L/P&S/5/10: letter no. 122, Auckland to Secret Committee, Fort Williams, November 13, 1840.

49. Graham, *China Station*, p. 140 n. 3. See also Bernard Brodie, *Sea Power in the Machine Age: Major Naval Inventions and Their Consequences on International Politics, 1814–1940* (London, 1943), pp. 156–57; Tramond and Reussner, pp. 52–54; and Preston and Major, p. 6.

50. Peacock, *Biographical Notes*, pp. 28–29.

51. Hoskins, p. 261.

52. During the Crimean War, Macgregor Laird wrote a scathing attack on official resistance to iron boats; see Cerberus (pseud.), "Somerset House Stops the Way," *Spectator* 27 (September 9, 1854).

# CHAPTER TWO

# The *Nemesis* in China

For several centuries, China and Europe had coexisted at arm's length, having only limited contact. Each side knew a great deal about the other. The Chinese imported European clocks and instruments, and respected Western astronomy and mathematics. Europeans, in turn, purchased the silks, porcelain, tea, and objets d'art of China, and admired some of her customs and institutions.

Yet the exchange was quite restricted and showed little promise of growth, even after two and a half centuries of direct contact. This was especially galling to the British, who by the eighteenth century were the dominant European power in the Far East and had developed a national craving for Chinese tea. In exchange for this tea, Britain had little to offer, for China was economically self-sufficient. Hence the tea trade caused a serious drain of gold and silver to China. The Chinese government, interested mainly in keeping the "sea barbarians" under control, deliberately confined the trade to certain merchants of Canton and resisted the entreaties of such distinguished ambassadors as Lord Macartney (1793) and Lord Amherst (1816).

The situation began to change, however, at the end of the eighteenth century, when the British discovered that their Indian possessions could produce a commodity for which China had a large and fast-growing demand: opium. As a result, there arose a triangular trade; India produced the opium, the Chinese exchanged the opium for tea, and the British drank the tea. From the British side of the triangle, there was not so much a flow of goods as of political and military power.

Opium, as Michael Greenberg pointed out, "was no hole-in-the-corner petty smuggling trade, but probably the largest commerce of the time in any single commodity."[1] At the center of this commerce stood the Honourable East India Company. The opium was grown mostly on company lands in Bengal, and though private merchants carried out the trade to China, they had to purchase their ware from the company, which was the sole opium manufacturer after 1797. Furthermore, the company's main business, outside of administering and taxing Indians, was the export of Chinese tea to England.

The company's monopoly of the China trade ended in 1834, and relations with China were soon exacerbated by the activities of private British traders. The Chinese government attempted several times to curtail the drug traffic, but with little success. What Chinese officials considered law enforcement against smuggling and narcotics, the traders saw as unjustified interference with free enterprise. William Jardine of the trading firm Jardine Matheson and Co. wrote anonymously to the *China Repository* in 1834:

> Nor indeed should our valuable commerce and revenue both in India and Great Britain be permitted to remain subject to a caprice, which a few gunboats laid alongside this city would overrule by the discharge of a few mortars. . . . The result of a war with the Chinese could not be doubted.[2]

That same year Jardine, Matheson, and sixty-two other British merchants in China petitioned the king to send three warships and a plenipotentiary to China, and "expressed the opinion that there would be no difficulty in intercepting the greater part of the internal and external trade of China and the capture of all the armed vessels of the empire.[3]

Even after the loss of its trade monopoly, the East India Company kept an interest in China through its import of tea and especially its manufacture of opium. Indeed opium yielded one seventh of the total revenue of British India in the nineteenth century. The China trade was essential to the prosperity of the British Empire. It is no surprise, then, that the East India Company became involved in the war with China.

If the tension between China and Britain was commercial in origin, its persistence was a consequence of the state of military technology. Like an elephant and a whale, China and Britain evolved in two different habitats. At sea, Britain was invincible and could destroy any Chinese fleet or coastal fort. China, on the other hand, was a land empire with few interests beyond her shores and few cities along her coasts. As long as the Europeans were incapable of pushing their way inland, the Celestial Empire was invulnerable.

The steamer, with its ability to navigate upriver and attack inland towns, ended the long Anglo-Chinese stalemate.

The first steamer to reach China, in 1830, was the 302-ton *Forbes,* a seagoing ship built in India for the Calcutta-Macao trade. In 1835, Jardine purchased a 115-ton steamer with two 24-horsepower engines, which he named *Jardine.* He and other foreign merchants then petitioned the senior Hong merchant, their Chinese counterpart, for permission to operate the *Jardine* on the Pearl River between Canton and Macao. The acting governor-general of Canton rejected their request:

> . . . if he presumes obstinately to disobey, I, the acting governor, have already issued orders to all the forts that when the steamship arrives they are to open a thundering fire and attack her. On the whole, since he has arrived within the boundaries of the Celestial Dynasty, it is right that he should obey the laws

of the Celestial Dynasty. I order the said foreigner to ponder this well and act in trembling obedience thereto.

When the *Jardine* steamed upriver in defiance of this order, she was fired on and forced to retreat.[4]

This setback notwithstanding, the foreigners, who now possessed steamers, refused to submit to threats of thundering fire. The British merchants continued to petition their government to send a punitive expedition to China. The British government, in particular the redoubtable Palmerston, also believed a war with China was inevitable. The only question was how to carry it out. Jardine, writing to Palmerston in November 1839, advised sending two line-of-battle ships, four frigates, two or three sloops, two large steamers, and "two small flat bottomed Steamers for River work, which it may be necessary to take out in frame and set up in China."[5] Palmerston, who hoped to bring China to her knees by seizing an off-shore island and stopping her coastal trade, passed on this advice to the lords of the Admiralty.[6] Sir Gilbert Elliott, Lord Minto, at the time first lord of the Admiralty, had a clearer idea of the difficulties involved. On February 16, 1840, he wrote to his nephew Lord Auckland:

I hope you will be able to send a respectable land force with the expedition. The mere occupation of an Island would not require much, but I think it very probable that the possession of one or two of their towns or great commercial Depots on the line of inland communication, but which are approachable from the sea may be very desirable, and to effect this a considerable force in troops would be necessary. However you are accustomed to work upon so great a scale that I feel no apprehension of your stinting this operation of whatever you think likely to secure and expedite its' success and shall not be much surprised if I receive a letter from Emily dated the Imperial Palace of Pekin. For after all it is nothing more nor less than the conquest of China that we have undertaken. I believe I have already told you that we turn to the Indian Government for such steamers as it may be able to furnish, we have none fit for such a voyage except some giants of great draught of water, which would be of little use in inshore and river operations, and consume an enormous quantity of fuel.[7]

The parties involved in planning the war with China—Jardine, Palmerston, Hobhouse, Minto, and Auckland—relied mainly on the traditional tools of war (that is, sailing warships and marine infantry) to defeat their enemy. If they thought of river steamers, it was at most as small auxiliary vessels that would have to be assembled in China, a lengthy and complicated process. Peacock, however, had other plans, as John Laird explained:

> The China war having commenced it was decided by the Secret Committee of the East India Co, on the recommendation of Mr Peacock, instead of sending all these Vessels out in pieces, and putting them together at Bombay, to send 4 of them under Steam round the Cape, an experiment at that time considered very hazardous, especially as the "Nemesis" and "Phlegethon" (two Vessels built by Mr Laird at Birkenhead) had to carry two 32 pounders on pivots, one at each end of the Vessel.[8]

We know a great deal about the *Nemesis* as a machine and as a protagonist, thanks to two sources. One is a report to the Admiralty by the naval architect Augustin Creuze, who at the Royal Navy Dockyard at Portsmouth examined the ship after she hit a rock off the Bay of St. Ives on the Cornish coast.[9] The other was a book based on the notes of Captain William Hutcheon Hall, who recounted the ship's history after she left England.[10]

Creuze gave a detailed technical description of the *Nemesis*. She measured 660 tons burthen, 184 feet long overall (165 between perpendiculars), 29 feet wide, 11 deep with a draft of 6 feet when fully loaded and less than 5 in battle trim. She was powered by two 60-horsepower Forrester engines and had two masts. The armament she carried was heavy for a boat her size: two pivot-mounted 32-pounder guns powerful enough to blast a hole in a fortress wall. In addition, she had five brass 6-pounders, ten smaller cannons, and a rocket launcher. Her interior was separated by bulkheads into seven watertight compartments. She also had two sliding keels and an adjustable rudder. During trials she tore a gash in her hull; the safety offered by her bulkheads and the ease with which she was repaired impressed Creuze.

He also discussed the inaccuracy of compasses, which had bedeviled iron ships from their inception. An iron hull deflects the earth's magnetism, making uncorrected compasses unreliable on board. In 1838 the Astronomer-Royal, Professor George Airy, discovered a method of compensating for the influence of the hull, and applied it to several ships. Although he did not adjust the compass of the *Nemesis* himself, the ship was deemed fit to go to sea.[11] Creuze concluded, in his report to the Admiralty, that iron was a better material for shipbuilding than timber.

To command the *Nemesis* the Secret Committee chose Hall, a master in the Royal Navy. As a young man, Hall had accompanied Lord Amherst's 1816 mission to China. In the late 1830s he became interested in steamers and spent two years studying them in Glasgow and along the Clyde and the Mersey. In June 1839 he crossed the Atlantic on the *British Queen,* one of the first steamships belonging to Macgregor Laird's British and American Steam Navigation Co. While in America, he paid close attention to the steamboats on the Hudson and Delaware rivers, then returned to England just as the Birkenhead Iron Works was putting the final touches on the *Nemesis.* Thus prepared, Hall joined the *Nemesis* in December 1839. On February 14, 1840, Peacock officially requested the Admiralty's permission to appoint him captain of the ship. The request was granted on February 26.[12]

The *Nemesis* left Portsmouth on March 28, 1840. Her journey east was long and difficult. She was the first iron ship to round the Cape of Good Hope and almost sank in the Indian

Ocean when heavy seas caused her hull to split open next to the paddle-boxes. After some improvised repairs at Delagoa Bay, she steamed for Ceylon, arriving October 6. There Hall received orders to proceed to China. Peacock meanwhile was exultant: "I am in high spirits about my iron chickens; having excellent accounts of them from Maderia [*sic*]. I have accounts of 'Nemesis' from the Cape, where she arrived in fine order, and literally astonished the natives."[13]

By the time the *Nemesis* reached Macao on November 25, 1840, the war had been going on, desultorily, for five months. The British fleet had harassed coastal towns like Amoy and made preparations for an offensive against Canton. On January 7, 1841, strengthened by the arrival of the *Nemesis* and some seagoing steamers, the British launched their first attack on the Bogue forts defending the Pearl River below Canton. The Chinese, whose defense strategy was static, had hoped to hold off their enemy, but the broadsides of the British men-of-war, towed into position by the steamers, quickly breached their defenses. Marine infantry troops soon stormed the fortifications.

The Chinese fleet was equally vulnerable. The war junks were half the size of the *Nemesis*, or one tenth that of a first-rate British battleship. They were armed with small cannons that were hard to aim, and with boarding nets, pots of burning pitch, and handguns. Without much difficulty the *Nemesis* sank or captured several junks; the rest were frightened off with Congreve rockets. The Chinese also relied on fire-rafts filled with gunpowder and oil-soaked cotton that were set ablaze and pushed toward enemy ships. The steamers, however, quickly grappled them and towed them out of reach. The previous year Commissioner Lin Tse-Hsü had purchased the 1080-ton American merchantman *Cambridge,* but for lack of sailors who knew how to handle the ship, she was kept idle behind a barrier of rafts. She was soon lost to the *Nemesis.*[14]

In a few days the river route to Canton was clear and the

sailing fleet began its slow ascent. As the fleet approached the city in early February, the *Nemesis* entered the inner passage, a labyrinth of narrow shallow channels that paralleled the main channel of the river, a place where no foreign warship had ever dared venture. While approaching Canton from the rear, she destroyed forts and junks at will and terrorized the inhabitants. Commodore J. J. Gordon Bremer, commander in chief of the Expedition Fleet, described the role of the *Nemesis* in a letter to the earl of Auckland:

> On proceeding up to Whampoa, three more dismantled forts were observed, and at four P.M. the Nemesis came into that anchorage having (in conjunction with the boats) destroyed five forts, one battery, two military stations, and nine war junks, in which were one hundred and fifteen guns and eight ginjalls, thus proving to the enemy that the British flag can be displayed throughout their inner waters wherever and whenever it is thought proper by us, against any defence or mode they may adopt to prevent it.[15]

Hall was exultant. On March 30, 1841, he wrote to John Laird:

> It is with great pleasure I inform you that your noble vessel is as much admired by our own countrymen as she is dreaded by the Chinese, well may the latter offer a reward of 50,000 dollars for her . . . but she will be difficult to take, they call her the devil ship and say that our shells and Rockets could only be invented by the latter. They are more afraid of her than all the Line-of-Battle ships put together.[16]

And two months later he wrote Peacock:

> With respect to the Nemesis I cannot speak too highly in her praise she does the whole of the advanced work for the Expedition and what with towing Transports, Frigates, large Junks, and carrying Cargoes of provisions, troops and Sailors, and repeatedly coming in contact with Sunken Junks—Rocks, Sand banks, and fishing Stakes in these unknown waters, which we are obliged to navigate by night as well as by day, she must be the strongest of the strong to stand it. . . . as far as fighting goes we

have had enough of that being always in advance, and most justly do the Officers as well as the Merchants of Macao say "that she is worth her weight in gold."[17]

Technically the British campaign of 1841 was a huge success, and Hall may be forgiven for his enthusiasm over the performance of his boat. But the political results of the campaign were nonetheless disappointing. The British fully expected the government of China to sue for peace following the destruction of the Chinese fleet and the capture of the Bogue forts and of Canton. The Chinese, though, were persuaded neither by these victories nor by the subsequent British capture of Amoy, Tinghai, Chinghai, and Ningpo. The British commander in chief, Admiral George Elliott, therefore decided to strike at the Grand Canal—the jugular vein of China—the principal north-south trade route along which boatloads of rice from Szechuan province were sent to feed the population of Peking, the capital. The idea may have originated in a letter which Samuel Baker, tea inspector for the East India Company, wrote in February 1840 and which was transmitted to Palmerston:

> The Yang coo kiang as far as its junction with the Grand Canal ought to be examined and regularly surveyed. This might be done with the aid of a steamer. . . . The island of Kiu Shan would be a strong position and enable us to distress the internal commerce greatly by cutting off the communication between the Northern and Southern Provinces by means of the Grand Canal.[18]

Palmerston reiterated the proposal in a secret instruction to the lords of the Admiralty dated February 20, 1840.[19]

By the start of the 1842 campaign, the British fleet had been reinforced by several additional steamers. The Indian Navy sent its steamships *Atalanta, Madagascar, Queen,* and *Sesostris.* The gunboats destined for the Indus also appeared on the scene: the 510-ton *Phlegethon,* sister of the *Nemesis,* and the smaller *Medusa, Ariadne, Pluto,* and *Proserpine.* Altogether the fleet that advanced upon the Yangtze in May 1842 included eight sailing warships, ten steamers, and over fifty transports, troopships, and schooners.[20]

The Chinese had a vague idea of how steamers worked. Commissioner Lin called them "wheeled vehicles which use the heads of flames to drive machines, cruising very fast," and Ch'i-shan, governor-general of Chihli, called them "ships with wind-mills" and "fire-wheel boats."[21] An anonymous Chinese source, whose translated account of steamers appeared in *The Nautical Magazine,* described the *Nemesis* in these terms:

> On each side is a wheel, which by the use of coal fire is made to revolve as fast as a running horse. . . . Steam Vessels are a wonderful invention of foreigners, and are calculated to offer delight to many.[22]

In June 1842 the British fleet entered the Yangtze. The Chinese were ready to receive their enemy, having assembled a considerable fleet of sixteen war junks and seventy merchantmen and fishing vessels requisitioned for naval duty. In the forts of Woosung, near the mouth of the river, they had placed 253 heavy artillery pieces.

The Chinese also unveiled a secret weapon: paddle-wheelers armed with brass guns, gingals, and matchlocks, and propelled by men inside the hull operating treadles. Nin Chien, governor-general of Nanking, wrote of them:

> Skilled artisans have also constructed four water-wheel boats, on which we have mounted guns. They are fast and we have specially assigned Major Liu Ch'ang . . . to command them. If the barbarians should sail into the inland waterways, these vessels can resist them. There is not the slightest worry.[23]

The battle of Woosung was swift. The British ships of the line soon silenced the guns of the forts. The *Nemesis,* towing the eighteen-gun *Modeste,* led the fleet into the river, firing grape and canister at the Chinese crafts, which fled. The *Nemesis* and the *Phlegethon* thereupon chased the fleeing boats, captured one junk and three paddle-wheelers, and set the rest on fire.

The British were astonished to discover that their opponents had paddle-wheelers. Some saw them as proof of the Chinese imitative ability.[24] In actuality the Chinese took the idea of paddle-wheels from their own history—the paddle-wheel boat was a Chinese invention of the eighth century or earlier. Under the Sung dynasty, paddle-wheelers played a celebrated role in battles against pirates in 1132 and against the armies of Digunai in 1161. From these examples, the hard-pressed Nin Chien and other Chinese officials drew inspiration in 1841 and 1842.[25]

Even more ironic is the appearance of the paddle-wheel in the West. The first paddle-wheel steamer was built in 1788 by Patrick Miller and William Symington, after Miller recalled having read somewhere "that the Chinese had, in the long-distant past, tried paddle-wheels fitted to certain of their junks with the cranks turned by slaves. . . ."[26] Like so many other Chinese inventions, the paddle-wheel was to haunt China in later centuries when her innovative spirit had flagged and her technology was surpassed by that of the Western barbarians.

After the one-sided battle of Woosung, the British fleet encountered little resistance from the Chinese. Instead, its lumbering journey upriver was marked by a constant struggle against currents, sandbars, and mud. Every one of the sailing ships had to be towed by the steamers again and again. Finally, in July 1842, the fleet reached Chinkiang, at the intersection of the river and the Grand Canal. This time the court at Peking realized its precarious situation and a few days later sent a mission to Nanking to sign a peace treaty. Steam had carried British naval might into the very heart of China and led to the defeat of the Celestial Empire.

After the Opium War, small armed steamers continued to serve in the Far East. The *Nemesis* was assigned to chase pirates in the Philippine and Indonesian archipelagos. In the Second Anglo-Burmese War of 1852–53, the British advanced up the Irrawaddy with an entire fleet of steamers, many of them veterans of the Opium War. During the Second Opium War (1856–60), the Royal Navy brought up more than twenty-five gunboats and other small steamers to attack Canton and the Taku forts near Peking. Gunboats also figured prominently in the French conquests of Tonkin (1873–74) and Annam (1883), and in the Third Anglo-Burmese War of 1885.[27]

The gunboat had become not just the instrument, but the very symbol of Western power along the coasts and up the navigable rivers of Asia. One protagonist of the colonial wars of that time, Colonel W. F. B. Laurie, put it succinctly: Steamers, he declared, were "a 'political persuader,' with fearful instruments of speech, in an age of progress!"[28]

The early history of the gunboat illustrates the interaction between technological innovation and the motives of imperialism. Because his father had an iron foundry and a shipyard, Macgregor Laird could turn his interest in Africa into an exploring expedition on the Niger. Peacock's classical erudition and Russophobia led him to translate the Anglo-Indian concerns with rapid communications into a steamboat expedition on the Euphrates. Their combination of interests persuaded the East India Company to become the first major purchaser of gunboats. In turn, the company's habit of acquiring gunboats led to Britain's victory in the Opium War. Thus, in the case of gunboats, we cannot claim that technological innovation caused imperialism, nor that imperialist motives led to technological innovation. Rather, the means and the motives stimulated one another in a relationship of positive mutual feedback.

## NOTES

1. Michael Greenberg, *British Trade and the Opening of China, 1800–42* (Cambridge, 1951), p. 104.

2. K. M. Panikkar, *Asia and Western Dominance* (New York, 1969), p. 97.

3. William Conrad Costin, *Great Britain and China, 1833–60* (Oxford, 1937), p. 27.

4. On the *Forbes* and the *Jardine*, see George Henry Preble, *A Chronological History of the Origin and Development of Steam Navigation*, 2nd ed. (Philadelphia, 1895), pp. 142–45; H. A. Gibson–Hill, "The Steamers Employed in Asian Waters, 1819–39," *Journal of the Royal Asiatic Society, Malayan Branch*, 27 pt. 1 (May 1954):127 and 153–56; H. Moyse–Bartlett, *A History of the Merchant Navy* (London, 1937), p. 229; Arthur Waley, *The Opium War Through Chinese Eyes* (London, 1958), pp. 105–06; and Peter Ward Fay, *The Opium War, 1840–1842: Barbarians in the Celestial Empire in the Early Part of the Nineteenth Century and the War by Which They Forced Her Gates Ajar* (Chapel Hill, N.C., 1975), p. 51.

5. India Office Records, L/P&S/9/1, pp. 411–12.

6. India Office Records, L/P&S/9/1, pp. 487–88.

7. National Maritime Museum, Greenwich, ELL 234: Letters from Sir Gilbert Elliot, Lord Minto, on China 1839–41.

8. John Laird, "Memorandum as to the part taken by the late Thomas Love Peacock Esq in promoting Steam Navigation" (1873), MS Peacockana 2 in The Carl H. Pforzheimer Library, New York.

9. "On the Nemesis Private Armed Steamer, and on the Comparative Efficiency of Iron-Built and Timber-Built Ships," *The United Service Journal and Naval and Military Magazine*, part 2 (May 1840):90–100.

10. This book went through three editions: The first two are William Dallas Bernard, *Narrative of the Voyages and Services of the Nemesis from 1840 to 1843*, 2 vols. (London, 1844 and 1845); the third edition is Captain William H. Hall (R.N.) and William Dallas Bernard, *The Nemesis in China, Comprising a History of the Late War in That Country, with a Complete Account of the Colony of Hong Kong* (London, 1846). The book was reviewed in "Voyages of the 'Nemesis'," *The Asiatic Journal and Monthly Miscellany* 3, 3rd series (May–Oct. 1844):355–59. For a recent look at the subject, see

David K. Brown, "Nemesis, The First Iron Warship," *Warship* (London) 8(Oct. 1978):283–85.

11. "Mr. Airy, Astronomer-Royal, on the Correction of the Compass in Iron-Built Ships," *The United Service Journal and Naval and Military Magazine* part 2 (June 1840):239–41. See also Stanislas Charles Henri Laurent Dupuy de Lôme, *Mémoire sur la construction des bâtiments en fer, adressé à M. le ministre de la marine et des colonies* (Paris, 1844), pp. 36–41; Edgar C. Smith, *A Short History of Naval and Marine Engineering* (Cambridge, 1938), pp. 99–100; and Hall and Bernard, p. 4.

12. India Office Records, L/P&S/3/6, p. 167. On Hall's career, see "William Hutcheon Hall," in William R. O'Byrne, *A Naval Biographical Dictionary: Comprising the Life and Services of Every Living Office in Her Majesty's Navy, from the Rank of Admiral of the Fleet to that of Lieutenant, Inclusive* (London, 1849), pp. 444–46; and "Hall, Sir William Hutcheon," in *Dictionary of National Biography*, 8:978.

13. Edith Nicolls, "A Biographical Notice of Thomas Love Peacock, by his Granddaughter," in Henry Cole, ed., *The Works of Thomas Love Peacock*, 3 vols. (London, 1875), 1:xliii. On the eastward journey of the Nemesis, see Hall and Bernard, ch. 1.

14. On Chinese defenses, see John Lang Rawlinson, *China's Struggle for Naval Development, 1839–1895* (Cambridge, Mass., 1967), pp. 3–5 and 16–18; Fay, pp. 123–24, 207–09, 218, and 289; and G. R. G. Worcester, "The Chinese War Junk," *Mariner's Mirror* 34(1948):22.

15. Reprinted in *The London Gazette Extraordinary* 19984 (June 3, 1841):1428.

16. India Office Records, L/MAR/C 593, pp. 543–44.

17. India Office Records, L/P&S/9/7, pp. 59–60.

18. India Office Records, L/P&S/9/1, p. 519.

19. India Office Records, L/P&S/9/1, p. 591.

20. The Yangtze campaign of 1842 is described in Gerald S. Graham, *The China Station: War and Diplomacy 1830–1860* (Oxford, 1978), ch. 8; G. R. G. Worcester, "The First Naval Expedition on the Yangtze River, 1842," *Mariner's Mirror* 36(1950):2–11; Hall and Bernard, pp. 326–27; Rawlinson, pp. 19–21; and Fay, pp. 313 and 341–45.

21. Lo Jung-Pang, "China's Paddle-Wheel Boats: Mechanized Craft Used in the Opium War and Their Historical Background," *Tsinghua Journal of Chinese Studies*, NS no. 2 (1960):190–91. For a different translation of Lin's description, see Waley, p. 105.

22. Letter from William Huttman, *The Nautical Magazine and Chronicle* 12(1843):346.

23. Lo, p. 190.

24. Lo, p. 194; Worcester, "War Junk," pp. 23–24; and Fay, p. 350.

25. Lo, pp. 194–200; and Rawlinson, pp. 19–21.

26. George Gibbard Jackson, *The Ship Under Steam* (New York, 1928), p. 26.

27. On the last assignment of the *Nemesis*, see *Parliamentary Papers* 1851 (378.) vol. LVI part 1, pp. 149–52. On the Second Anglo-Burmese War, see Col. W. F. B. Laurie, *Our Burmese Wars and Relations with Burma: Being an Abstract of Military and Political Operations, 1824–25–26, and 1852–53* (London, 1880), pp. 86–92. On gunboats in the Second Opium War, see Antony Preston and John Major, *Send a Gunboat! A Study of the Gunboat and its Role in British History, 1854–1904* (London, 1967), ch. 4. On French gunboats in Indochina, see Joannès Tramond and André Reussner, *Eléments d'histoire maritime et coloniale (1815–1914)* (Paris, 1924), pp. 344–49, and Frédérick Nolte, *L'Europe militaire et diplomatique au dix-neuvième siècle 1815–1884*, vol. 3: *Guerres coloniales et expéditions d'outre-mer 1830–1884* (Paris, 1884), p. 521. On the Third Anglo-Burmese War, see A. T. Q. Stewart, *The Pagoda War: Lord Dufferin and the Fall of the Kingdom of Ava, 1885–6* (London, 1972), ch. 5.

28. Laurie, p. 109.

CHAPTER THREE

---

# Malaria, Quinine,
# and the Penetration
# of Africa

By the time Columbus first sighted the Americas, the Portuguese were well acquainted with the west coast of Africa, for they had been exploring it for sixty years. Yet, during the next three and a half centuries, Africa remained in the eyes of Europeans the "dark continent," its interior a blank on their maps, as they chose instead to explore, conquer, and settle parts of the Americas, Asia, and Australia.

How can we explain this paradox? For one thing, there was little motivation for Europeans to penetrate Africa before the nineteenth century. The slave traders—Africans and Europeans alike—who met along the coasts to conduct their business wanted no outsiders with prying eyes disrupting their operations. Furthermore, despite legends of fabulous wealth, there was little concrete evidence that the profits to be derived from the penetration of Africa would even approximate those resulting from the slave trade or from trade with Asia and the Americas. Thus, the penetration of Africa that occurred in the nineteenth century was tied closely to missionary and abolitionist movements reacting against the slave trade.

But even more significantly, the means of penetration were

also lacking. Much of Africa is a plateau. Rivers cascade from the highlands to the sea in a series of cataracts. The coasts are lined with mangrove swamps and sandbars. And throughout the tropical regions, pack animals could not survive the nagana or animal trypanosomiasis. Those who wished to enter Africa would have to do so on foot or in dugout canoes.

These deterrents were by no means absolute prohibitions. After all, Europeans had explored the Americas with primitive means of transportation, despite difficult climates and topographies. It was disease that kept Europeans out of the interior of Africa. Although steamboats came to Africa and Asia at the same time, in Asia they wrought a revolution in the power of Europeans, whereas in Africa their effect was postponed for several decades. Before Europeans could break into the African interior successfully, they required another technological advance, a triumph over disease.

In his novel *War of the Worlds,* H. G. Wells described a group of extraterrestrial creatures who invade the earth in strange futuristic vehicles. As they are about to take over the globe, they are decimated by invisible microbes and are forced to flee. Wells could just as well have been writing about the various European attempts to penetrate Africa before the middle of the nineteenth century. In 1485 the Portuguese captain Diogo Cão sent a party of men to explore the Congo River; within a few days so many had died that the mission had to be called off. In 1569, Francisco Barreto led an expedition up the Zambezi valley to establish contact with the kingdom of Monomotapa; 120 miles upriver, the horses and cattle fell victim to trypanosomiasis and the men succumbed to malaria. Henceforth until 1835, Portuguese communications with the Zambezi interior were carried on through African or part-African agents.[1]

Similarly, in 1777–79, during William Bolts' expedition at Delagoa Bay, 132 out of 152 Europeans on the journey died. Mungo Park's 1805 venture to the upper Niger resulted in the death of all the Europeans present. In 1816–17, Captain James Tuckey led an exploring party up the Congo River, in which 19 out of 54 Europeans perished.[2]

These setbacks in no way curtailed European attempts to explore Africa. Each generation spawned a fresh crop of adventurers willing to risk their lives to investigate the unknown continent. With the nineteenth century appeared new motives to do so: a revival of the Christian proselytizing spirit, the abolition of the Atlantic slave trade and a curiosity elevated to the rank of scientific research and funded by a newly wealthy bourgeoisie. Among the enterprising explorers of this era was Macgregor Laird, the younger son of the shipbuilder William Laird, who was to play a pioneering role in opening up Nigeria to British influence. In the early 1830s, his father's firm had just begun to build iron steamboats. Macgregor Laird, then twenty-three years of age, was not content to remain the junior partner in a struggling new business. He was imbued with that restless spirit—part missionary fervor, part scientific curiosity, part commercial hope—which inspired so many nineteenth-century Britons to venture out and remake the world. In 1832 he saw opportunity beckon along the Niger River. Three decades earlier Mungo Park had explored the upper reaches of this river down to the Bussa Rapids. Then in 1830 the brothers Richard and John Lander traveled north from Lagos to the rapids and sailed downriver in a canoe, thus proving that the Niger and the Oil rivers, which flowed into the Bight of Benin through a mangrove swamp, were one and the same. When the Landers returned to England with the tale of their discovery, Laird realized that a boat capable of sailing up the river with a cargo of trade goods—in other words, a steamer—would open up an immense part of Africa to the commerce and influence of Great Britain.[3] To do so, he later wrote, would please

> . . . those who look upon the opening of Central Africa to the enterprise and capital of British merchants as likely to create new and extensive markets for our manufactured goods, and

fresh sources whence to draw our supplies; and those who, view-
ing mankind as one great family, consider it their duty to raise
their fellow creatures from their present degraded, denational-
ized, and demoralized state, nearer to Him in whose image they
were created.[4]

In 1832, therefore, Macgregor Laird and several Liverpool
merchants founded the African Inland Commercial Company
"for the commercial development of the recent discoveries of
the brothers Lander on the River Niger." The directors sought
a charter and a subsidy from the treasury but were refused.
They went ahead with their venture anyway and hired Richard Lander to lead their expedition. They bought the brig
*Columbine* as a storeship and ordered two steamers in which
to ascend the Niger. The larger of the two, the *Quorra,* was
built of wood by Seddon and Langley. She measured 112 by 16
feet, drew 7 feet at sea and 5½ feet on the river; she was powered by a 40-horsepower engine and carried a 26-man crew.
Macgregor Laird himself built the smaller one, the *Alburkah.*
She was 70 feet long by 13 feet wide with a draft of 4 feet 9
inches. Except for the deck, she was made entirely of iron. She
had a 15-horsepower Fawcett and Preston engine and carried
a crew of 14. Both boats were heavily armed. In addition to
handguns, the *Quorra* had a 24-pound swivel gun, eight
4-pound carriage guns, and an 18-pound carronade. The *Alburkah* carried a 9-pounder and six 1-pounder swivel guns.[5]

Under Laird's command, the little fleet reached the Niger
delta without incident. Leaving the *Columbine* in the Bight
of Benin, the steamers then went upriver, past the trading
towns of the delta to the confluence of the Niger and the Benue. There Laird hoped to found a trading post and to buy
palm oil at low prices.

Laird's steamers succeeded admirably in their assigned task,
and for this Laird deserves his reputation as an innovator and
an explorer. As a cultural and commercial mission, however,
the expedition was a failure and his expectations were shattered. Of the forty-eight Europeans present on the trip, only

nine returned, the rest having died of disease. Laird himself returned ill to England in January 1834 and never quite recovered his health. Despite the power of steam, the African environment had once again defeated the European attempt at penetration.[6]

Though very few Europeans ventured into the interior of Africa before the mid-nineteenth century, a substantial number had for centuries been trading along the coasts. After 1807, in an attempt to end the slave trade, the British government stationed a fleet along the West African coast to intercept slaving ships. Small army units were also placed at intervals along the shores to lend weight to the abolition campaign. Here and there the first Christian missions were founded. These various groups of whites were subject to the diseases prevailing in the region.

We know much more about the death rates among British military personnel in West Africa than among the slave traders, their predecessors, for this was the time when the keeping of statistical records became a vital part of Western culture. The Royal African Corps, stationed from the Gambia to the Gold Coast, was composed of military criminals and offenders allowed to exchange their sentences for service in Africa. In most cases this meant substituting death for prison. In 1840 the *United Service Journal and Naval and Military Magazine* devoted an article to the health of these troops.[7] It gave the following figures. Of the 1,843 European soldiers who served in Sierra Leone between 1819 and 1836, 890, or 48.3 percent, died. The worst year was 1825, in which 447 out of 571 (78.3 percent) succumbed to disease. Despite a constant influx of European arrivals, the size of the garrison declined by over a hundred each year. The Gold Coast was just as deadly: Two-thirds of the Europeans who landed there in the years 1823–27 never lived to return home; in the year 1824 alone, 221 out of 224 lost their lives. On the whole, 77 percent of the white soldiers sent to West Africa perished, 21 percent became invalids, and only 2 percent were ultimately found fit for future service.

Among West Indian soldiers stationed in the same region, the death rate was only one tenth that for whites, though still twice that prevailing in their native lands. During the 1825–26 epidemic in the Gambia which killed 276 out of 399 whites, only one out of 40 or 50 West Indians fell victim to the illness. It is likely that the epidemic in question was yellow fever, a disease endemic to the West Indies against which many West Indians had developed a resistance. In 1830 the British government recognized the significance of the death rates and stopped sending white troops to West Africa, except for half a dozen sergeants to command the West Indian soldiers.

The authors of the article, of course, did not understand the exact causes of this horrendous situation. At least they did not blame the men themselves, for they noted that robust, teetotalling English missionaries living on the same coast were as likely to suffer the effects of the disease; of 89 who went to West Africa between 1804 and 1825, 54 died and another 14 returned in bad health. Nor was the climate to blame, for dry and windy stations were as dangerous as those adjoining fetid marshes. The cause of the problem, they concluded, was fevers, either yellow or remittent. A scientific approach was beginning to replace the moralistic judgments of former times.

Philip Curtin, in his writings on the question, gives equally appalling death rates. Among British military personnel recruited in the United Kingdom who served during the years 1817 to 1836, the death rates per thousand were:

| | |
|---|---|
| on the eastern frontier of South Africa (1817–36) | 12.0 |
| in the United Kingdom (1830–36) | 15.3 |
| in Tenasserim, Burma (1827–36) | 44.7 |
| in Ceylon (1817–36) | 75.0 |
| in Sierra Leone (1817–36) (deaths from diseases only) | 483.0 |
| in Cape Coast Command, Gold Coast (1817–36) | 668.3 |

Among Europeans serving with the African Squadron of the Royal Navy off the coast of West Africa, the death rate in 1825–45 was 65 per thousand; among British troops in West Africa in 1819–36 it was 483 per thousand for enlisted men and 209 for officers. Meanwhile, West African soldiers serving in the British army in the same area suffered a death rate of only 2.5 per thousand. It is for this reason that Africa became known as the "white man's grave."[8]

Though dysentery, yellow fever, typhoid, and other ills contributed to the high death rates, the principal killer of Europeans in Africa was malaria. Throughout history malaria has probably caused more human deaths than any other disease. It exists in several varieties. Tertian malaria, endemic throughout much of the world, is caused by the protozoan *Plasmodium vivax* and produces intermittent fevers and a general weakening of the body. Another variety, brought on by the *Plasmodium falciparum,* is endemic only to tropical Africa and is far deadlier. It is found not only in swamplands and rainforests, but also in the drier savannas. The body's resistance, gained from a successful bout with the disease, is temporary at best, and many Africans suffer repeated low-level attacks throughout their lives. To adult newcomers to Africa, who have not had the opportunity to build up a resistance, the disease is most often fatal.

Early nineteenth-century European medical opinion, influenced by the age-old association of malaria with swamps, blamed humid air and putrid smells for the disease; hence the French word *paludisme* (from the Latin word for swamp) and the Italian *mal'aria,* or bad air. The strangest theory of all was put forth by Macgregor Laird. In trying to explain the epidemic that had decimated his Niger expedition, he wrote Thomas Peacock in 1837:

> Captain Grant mentioned the possibility of getting firewood at Fernando Po, nothing can be more injurious both to the Vessel

64

and the Crew . . . to the Crew, as the miasmatic exhalations from it will infallibly produce fever and disease. I have had melancholy experience of the effects of wood taken on board & used as Firewood for the Engines on the Coast of Africa.[9]

Not until 1880 did a French scientist, Alphonse Laveran, discover the *Plasmodium* that invades the bloodstream; and only in 1897 was the vector of malaria, the *Anopheles* mosquito, identified by the British physician Ronald Ross and the Italian scientists Giovanni Batista Grassi and Amico Bignami.[10]

That the cause of malaria was not known to science until the end of the century did not prevent a remedy from emerging much earlier out of a long process of trial and error. Before our own century, technological advances often preceded a scientific explanation of the underlying natural phenomena; technological advances arising out of scientific discoveries were the exception. We should not think of technology as "applied science" before the end of the nineteenth century, but rather of science as "theoretical technology."

For centuries, people had sought relief from the dreaded disease. In the seventeenth century, Jesuits had introduced the bark of the cinchona tree as a cure for *vivax* malaria and disseminated it in Europe. Cinchona bark, though effective, had a number of drawbacks. Because it came from trees that grew only in the Andes, the supply in Europe was often limited. Making matters worse, what did reach the consumers was not only expensive but often adulterated or deteriorated. Moreover, its Jesuit connection made it suspect among Protestants; Oliver Cromwell, dying of malaria, is said to have refused the "popish" remedy. It also was useless against yellow fever and a number of other fevers that were then confused in medical theory. And finally, it had an awful taste.

Yet up through the eighteenth century, medical authorities regularly prescribed the bark. By the turn of the following century, though, physicians favored treating fevers with doses of mercury for salivation and calomel for its purgative quali-

ties. Frequent bleedings and blisters were other common treat-
ments. These "remedies" undoubtedly killed more patients
than they saved and must have contributed to the extraordinary death rates among British military personnel in West
Africa.[11]

The dawn of a breakthrough in treating malaria dates from
the year 1820, when two French chemists, Pierre Joseph Pel-
letier and Joseph Bienaimé Caventou, succeeded in extracting
the alkaloid of quinine from cinchona bark. Commercial production of quinine began in 1827, and by 1830 the drug was
being manufactured in large enough quantities for general
use.[12]

From the late 1820s on, doctors in malarial areas conducted
experiments with quinine and published the results of their
investigations. The first important experiments were carried
out in Algeria, following the French invasion of 1830. Serious
health problems plagued the French troops stationed there,
with typhoid and cholera outbreaks common occurrences. The
most severe problem, however, was malaria. Bône, which was
surrounded by swamps, had the highest incidence of disease in
Algeria, and epidemics broke out every summer. In 1832, of
the 2,788 French soldiers stationed in that town, 1,626 were
hospitalized. The next year, 4,000 out of 5,500 were similarly
affected, and out of every 7 hospitalized soldiers, 2 died. The
cause of these deaths was not disease alone, but also the treatment the patients received. French army doctors at the time
were influenced by Dr. J. Broussais, head of the army medical
school of Val-de-Grâce, who taught that fevers should be
treated with purgatives, bleedings, leeches, and a starvation
diet. Quinine, he believed, should be administered in tiny
doses only after the seventh or eighth attack; among other
reasons, the new drug was too expensive, at twenty-five francs
an ounce, for military use.

Two army physicians, Jean André Antonini and François
Clément Maillot, rebelled against the accepted practices of

their colleagues. Antonini noted that intermittent fevers responded to quinine, and this permitted him to distinguish malaria from typhoid fever. He moderated the bleedings and gave his patients more food. Maillot, posted to Bône at the height of the malaria epidemic of 1834, went further. At the first sign of fever, he prescribed twenty-four to forty grains of quinine immediately, instead of four to eight several days later as Broussais had taught. He also fed his patients a nutritious diet. The results were most impressive. Only one out of every twenty patients died, compared to two out of seven the year before. Consequently, sick soldiers began fleeing other hospitals to come to Maillot's. In 1835 he described his methods to the Académie de Médecine in Paris, and a year later he published his findings under the title *Traité des fièvres ou irritations cérébro-spinales intermittentes*. Yet it was many years before his methods were accepted by the French military medical service. Finally, toward the end of his life, Maillot was idolized as a hero of French science, and in 1881 the Scientific Congress of Algiers honored him with the phrase: "It is thanks to Maillot that Algeria has become a French land; it is he who closed and sealed forever this tomb of Christians."[13]

In West Africa, too, the use of quinine became more common, while purgings and bleedings gradually fell into disfavor. By the mid-1840s, Europeans in the Gold Coast regularly kept a jar of quinine pills by their bedside, to be taken at the first sign of chills or fever. Yet this treatment, although beneficial against the *vivax* form of malaria prevalent in Algeria, was generally insufficient against *falciparum* malaria. To defeat the *Plasmodium falciparum,* the human bloodstream had to be saturated with quinine before the onset of the first infection; in other words, throughout one's stay in *falciparum* areas, quinine had to be taken regularly as a prophylactic.

Two chance events led to this discovery. The first occurred in 1839, on board the *North Star* stationed off Sierra Leone.

While serving on the ship, twenty crew members took cinchona bark daily and one officer did not; he alone died of malaria. The second incident took place two years later, when the
British government sponsored the largest of all the Niger expeditions up to that time. With three new steamers—the 457-
ton *Albert* and *Wilberforce* and the 249-ton *Soudan*—Capt.
H. D. Trotter led 159 Europeans up the Niger to the confluence of the Benue. To avoid the health problems of previous
missions every known precaution was taken. The crew was
specially selected from among athletic young men of good
breeding, the ships were equipped with fans to dispel bad air,
and the expedition raced at top speed through the miasmic
delta to reach the drier climate of the upper river as soon as
possible. Nonetheless the first cases of fever appeared within
three weeks, forcing the *Wilberforce* and the *Soudan* to return
to the Atlantic as floating hospitals. Within two months,
forty-eight of the Europeans had died, and by the end of the
expedition another seven fell victim to the disease. Africa had
regained its terrible reputation among the British.[14]

Despite this disappointment, the Niger expedition of 1841
represents a major step toward a solution to the problem of
malaria, for the physician on board one of the ships, Dr.
T. R. H. Thomson, used the opportunity to experiment with
various drugs. Some crew members received cinchona bark
with wine, others got quinine; Dr. Thomson himself took
quinine regularly and stayed healthy. He later wrote his ob-
servations on the matter in an article entitled "On the Value of
Quinine in African Remittent Fever" which appeared in the
British medical journal *The Lancet* on February 28, 1846. A
year later, Dr. Alexander Bryson, an experienced naval physician, published his *Report on the Climate and Principal Diseases of the African Station* (London, 1847), in which he advocated quinine prophylaxis to Europeans in Africa. In 1848
the director-general of the Medical Department of the British
Army sent a circular to all British governors in West Africa,
recommending quinine prophylaxis.[15]

Yet quinine prophylaxis was not immediately adopted. It took a spectacular demonstration to achieve this end. In 1854, Macgregor Laird, never cured of his fascination with Africa, proposed still another expedition to that continent. Under contract with the Admiralty, he had a ship called the *Pleiad* specially built. She was a 220-ton iron propeller-steamer rigged as a schooner, designed to pull two or three barges behind her on her way up the Niger. As was usually the case, she was armed with a 12-pounder pivot gun, four smaller swivel cannons, rifles, and muskets. The crew consisted of twelve Europeans and fifty-four Africans.

Before the ship sailed, Dr. Alexander Bryson wrote a set of instructions in which he described the clothing, diet, activities, and moral influences best suited to protect the health of the crew. To prevent fevers he recommended that each crew member take six to eight grains of quinine a day from the time the ship crossed the bar until fourteen days after she returned to the ocean. The captain of the ship, Dr. William Baikie, was himself a physician and saw to it that the crew followed this advice. The *Pleiad* stayed 112 days on the Niger and Benue rivers, and returned with all the European crew members alive. Thomas Hutchinson, a member of the expedition, attributed this to Dr. Bryson's suggestions; as he put it,

> Since my first visit to Africa in 1850, I have felt firmly convinced—and that conviction urges me to impress my faith on all who read this work—that the climate would not be so fatal as it has hitherto proved to Europeans, if a different mode of daily living, a proper method of prophylactic hygiene, and another line of therapeutic practice in the treatment of fevers, were adopted. Before, and beyond all others, is the preventive influence of quinine as it was used in the "Pleiad," in the mode here described. . . .[16]

As the prophylactic use of quinine spread, and as purgings and bleedings vanished, the death rates fell significantly. Philip Curtin gives some statistics: In the Royal Navy's Africa Squadron, the mortality rate fell from 65 per 1,000 in 1825–45 to 22 per 1,000 in 1858–67; in 1874, during the two-month military expedition against Kumasi, only 50 of the 2,500 European soldiers died of disease; in 1881–97, among British officials in the Gold Coast, the rate was 76 per 1,000, and in Lagos it was 53 per 1,000. On the whole, the first-year death rates among Europeans in West Africa dropped from 250–750 per 1,000 to 50–100 per 1,000. To be sure, this was still five to ten times higher than the death rates for people in the same age bracket in Europe. Africa remained hostile to the health of Europeans. Yet psychologically the improvement was significant. No longer was tropical Africa the "white man's grave," fit for only the most ardent visionaries and the unluckiest recruits. It was now a place from which Europeans could reasonably hope to return alive. In Curtin's words, ". . . the improvement over the recent past was understood well enough in official and missionary circles to reduce sharply the most serious impediment to any African activity."[17]

One immediate consequence of quinine prophylaxis was a great increase in the number and success of European explorers in Africa after the mid-century. Exploration, of course, remained a dangerous business, but no longer was it quasi-suicidal. With the prospect before them of fruitful discoveries, perhaps even glory and wealth, many more adventurous souls volunteered in the service of knowledge. David Livingstone, the most lionized of all the explorers, first heard of quinine prophylaxis while he was in Bechuanaland in 1843. During his march across southern Africa in 1850–56 he took quinine daily. By 1857 he was convinced that quinine was a preventive. In preparation for his Zambezi expedition of 1858 he made his European crew take two grains of quinine in sherry every day. Throughout the expedition, many suffered from malaria, but only three out of twenty-five died. Later he came to doubt the efficacy of quinine as a preventive, for it only lessened the impact of the disease. His favorite remedy for malaria was a concoction of quinine, calomel, rhubarb, and resin of julep which he called "Livingstone Pills."[18]

In the footsteps of the explorers, lesser protagonists of European imperialism penetrated the African interior: missionaries, soldiers, traders, administrators, engineers, planters and their wives and children, and finally tourists. All of them needed their daily quinine. In India and other tropical areas, the influx of Europeans added to the growing demand for the drug.

Until the 1850s all the world's cinchona bark came from the forests of Peru, Bolivia, Ecuador, and Colombia, where the tress grew wild. As world demand increased, the bark exports of the Andean republics rose from two million pounds in 1860 to twenty million in 1881. At that point the Andean bark was swept from the world market by the competition of Indian and Indonesian bark, the result of deliberate efforts by Dutch and British interests.

The idea of growing cinchona in Asia had been discussed many times, but with little effect as long as demand was small. In the early 1850s, as demand grew, Dutch botanists and horticulturists in Java urged the Netherlands East Indies government to import cinchona seedlings. In 1853–54, Justus Charles Hasskarl, superintendent of the Buitenzorg Botanical Gardens in Java, traveled to the Andes under an assumed name and secretly collected seeds; most of them perished, however. In 1858–60, Clements Markham, a clerk at the India Office, aided by a gardener from the British Royal Botanic Gardens at Kew named Weir, traveled to Bolivia and Peru, again secretly, to collect seeds of the *Cinchona calisaya* tree. Simultaneously, the English botanist Richard Spruce and another Kew gardener, Robert Cross, collected 100,000 *C. succirubra* seeds and 637 young plants in Ecuador; of these, 463 seedlings reached India, forming the nucleus of the cinchona plantations at Ootacamund in the Nilgiri Hills near Madras.

There followed a period of intensive experimentation. At botanical gardens in Bengal, Ceylon, Madras, and Java, horticulturists and quinologists exchanged seeds and information, and provided cheap seedlings and free advice to planters. A hybrid species, *C. calisaya Ledgeriana,* grafted onto the stem of a *C. succirubra* tree, formed the basis of the Javanese cinchona plantations after 1874. Techniques such as mossing (cutting strips of bark and wrapping the trees in moss) and coppicing (cutting trees to the ground every six or seven years) greatly increased the yield of alkaloids. While Peruvian bark had a two percent sulphate of quinine content, scientific breeding in Java raised the content to six percent by 1900, and later to eight or nine percent.

In the late nineteenth century, after the demise of the Andean bark industry, a compromise was worked out between the British and the Dutch. Plantations in India produced a cheaper, less potent bark from which chemists extracted totaquine, a mixture of antimalarial alkaloids. Almost the entire Indian production was destined to British military and administrative personnel in the tropics, and the excess was sold in India. The Javanese industry, which produced the more potent and expensive pure quinine, captured over nine tenths of the world market by the early twentieth century. This world monopoly of cinchona resulted not only from scientific methods of cultivation, but also from a marketing cartel, the Kina Bureau of Amsterdam, which coordinated the purchase of bark and the price and quantity of quinine sold. Not until the Japanese conquest of Indonesia in World War Two and the development of synthetic malaria suppressants did this Dutch control over one of the world's most vital drugs come to an end.

Scientific cinchona production was an imperial technology par excellence. Without it European colonialism would have been almost impossible in Africa, and much costlier elsewhere in the tropics. At the same time, the development of this technology, combining the scientific expertise of several botanical gardens, the encouragement of the British and Dutch colonial governments, and the land and labor of the peoples of India and Indonesia, was clearly a consequence as well as a cause of the new imperialism.[19]

River steamers had overcome the obstacle of poor transportation, and quinine that of malaria. Together, they opened much of Africa to colonialism, that is, to the systematic intercourse with Europe on European terms. The scramble for Africa has often been explained as a consequence of French political psychology after the Franco-Prussian War, or of the ambitions of King Leopold II of Belgium, or as a byproduct of the Suez Canal. No doubt. But it was also the result of the combination of steamers, quinine prophylaxis, and, as we shall see, the quick-firing rifle. From among the myriad events of the scramble, let us consider only a few that illustrate the arrival of steamers on the rivers of Africa, their European crews now protected from a certain death by quinine prophylaxis.

Macgregor Laird had not sent expeditions up the Niger River out of curiosity or philanthropy alone. They were, in his eyes, investments that must surely pay off, for the Niger trade was both lucrative and necessary to Britain. Palm oil, which had replaced slaves as the principal export of southern Nigeria, was essential as the raw material for soap and as a lubricant for industrial machinery. But the price of palm oil was kept unreasonably high by the Niger delta middlemen who brought it to the coast, and by the small European traders who shipped it to Europe. The instrument that would break through these bottlenecks, Laird believed, was the steam engine. In 1851 he wrote Earl Grey that steam "will convert a most uncertain and precarious trade into a regular and steady one, diminish the risk of life, and free a large portion of the capital at present engaged in it. . . ."[20]

What was needed was a double application of steam. One was to be a steamship line between Britain and West Africa, which we shall consider in a later chapter. The other was a regular steamboat service along the Niger in order to bypass the Nigerian middlemen. Laird's first appeals were rejected. After the *Pleiad* expedition in 1854 had vindicated his faith, however, the Royal Geographical Society convinced the British government to support his projects. In 1857 the Foreign Office agreed to send Dr. Baikie to open relations with the Caliphate of Sokoto on the middle Niger. The Admiralty contracted with Laird to send three steamers up the Niger annually for five years.

The *Dayspring*, the *Rainbow*, and the *Sunbeam* were built by John Laird's Birkenhead shipyard for this service. In the course of their voyages they naturally aroused the resentment of the delta traders whose business they were ruining. In 1859, after traders attacked the *Rainbow* and killed two of her crew members, Laird appealed to the government for a warship to accompany his steamers. Two years later H.M.S. *Espoir* entered the Niger and destroyed the villages that had been responsible for the assault on the *Rainbow*. By the 1870s several British companies were trading on the Niger with armed steamers, and every year a military expedition steamed up the river to destroy any towns that resisted the British intrusion. By the 1880s, Sir George Goldie's United African Company, uniting all the trading interests in the area, kept a fleet of light gunboats patrolling the river year round. In 1885 the British government declared the Niger delta a protectorate. Despite sporadic resistance, no African town along the rivers and no war-canoe could withstand for long the power of British gunboats.[21]

The Niger River was the scene of the earliest and most active use of steamers by the invading Europeans, because it was the easiest to navigate in all of tropical Africa. The other major rivers—the Congo, the Zambezi, the upper Nile, and their tributaries—were broken by cataracts which barred access to them by seagoing steamers. Boats had to be brought in pieces, portaged around the rapids, and reassembled before they could be used to explore the upper reaches of these rivers. To portage the steamers and equipment for an entire expedition required labor, technology, organization, and financing on a scale that the Niger explorers had never faced.

Livingstone used a series of small steamers: the *Ma Roberts*, the first steel steamboat, on which he explored the Zambezi River up to the Kebrabasa Rapids in 1858; the *Pioneer*, in 1861; and the *Lady Nyassa*, which was carried in pieces around the falls to Lake Nyassa.[22] Samuel White Baker had the steamer *Khedive* transported to the upper Nile.[23] To open up the Congo river basin, Henry Stanley had a steamer, the nine-ton *En Avant*, carried in pieces from the Atlantic to Stanley Pool. Shortly thereafter Savorgnan de Brazza's *Ballay* also appeared on the Congo.

After that the number of steamers multiplied quickly, for exploration, conquest, trade, and missionary work. They were transported to the most remote regions of the continent. In 1895–97 the French lieutenant Gentil conquered the area of the Ubangi and Shari rivers and Lake Chad using the first aluminum steamer, the *Léon Blot*.[24] And in 1898 on his cross-Africa expedition, Commandant Marchand had two steamers and three rowboats carried from the Ubangi to the Nile, on which he then steamed to his celebrated confrontation with Kitchener at Fashoda.[25]

Given the harsh topography of much of Africa, and the lack of pack animals, it is doubtful whether Europeans could have penetrated so fast or dominated so thoroughly if they had had to go on foot. Regions lacking good water transportation—for example the Central Sudan, the Sahara, Ethiopia, and the Kalahari—were among the last to be colonized. The contrast between the ease of water transport and the difficulty of land transport in nineteenth-century Africa accounts in large part for the European patterns of penetration and control.

## NOTES

1. John Ford, *The Role of Trypanosomiasis in African Ecology: A Study of the Tsetse Fly Problem* (Oxford, 1971), p. 327.

2. For the death rates on exploring expeditions in Africa, see René-Jules Cornet, *Médecine et exploration: Premiers contacts de quelques explorateurs de l'Afrique centrale avec les maladies tropicales* (Brussels, 1970), p. 7; Philip D. Curtin, *The Image of Africa: British Ideas and Actions 1780–1850* (Madison, Wis., 1964), pp. 483–87 and " 'The White Man's Grave': Image and Reality, 1780–1850," *Journal of British Studies* 1(1961):105; and Michael Gelfand, *Rivers of Death in Africa* (London, 1964), p. 18 and *Livingstone the Doctor: His Life and Travels. A Study in Medical History* (Oxford, 1957), pp. 3–12.

3. Laird was not the first to think of using a steamer to penetrate Africa. Captain James Tuckey had intended to put a Boulton and Watt engine on his riverboat *Congo* to explore that river in 1816; the engine was too heavy for the boat, however, and had to be removed. See André Lederer, *Histoire de la navigation au Congo* (Tervuren, Belgium, 1965), p. 7; and John Fincham, *History of Naval Architecture* (London, 1851), p. 329.

4. Macgregor Laird and R. A. K. Oldfield, *Narrative of an Expedition into the Interior of Africa, by the River Niger, in the Steam-Vessels Quorra and Alburkah, in 1832, 1833, and 1834*, 2 vols. (London, 1837), 1:vi.

5. Laird and Oldfield, 1:5–9; Liverpool Shipping Register, Entry 92 and 93, 5 July 1832, cited in P. N. Davies, *The Trade Makers: Elder Dempster in West Africa, 1852–1972* (London, n.d.), p. 409 table 9; "Report from the Select Committee on Steam Navigation to India, with the Minutes of Evidence, Appendix and Index" in *Parliamentary Papers* 1834 (478.) XIV, p. 426.

6. On Laird's expedition, see Laird and Oldfield; K. Onwuka Dike, *Trade and Politics in the Niger Delta 1830–1885: An Introduction to the Economic and Political History of Nigeria* (Oxford, 1956), pp. 18 and 61–63; H. S. Goldsmith, "The River Niger; Macgregor Laird and Those Who Inspired Him," *Journal of the African*

*Society* 31 no. 85 (Oct. 1932): 383–93; Christopher Lloyd, *The Search for the Niger* (London, 1973), pp. 130–41; Sir Roderick Impey Murchison, "Address to the Royal Geographical Society of London," in *Journal of the Royal Geographical Society* 31 (1861):cxxvi–cxxviii; and C. W. Newbury, ed., *British Policy Toward West Africa* (Oxford, 1965), pp. 6 and 66–79.

7. "Western Africa and its Effects on the Health of Troops," *United Service Journal and Naval and Military Magazine* pt. 2 (Aug. 1840):509–19.

8. Philip Curtin, "Epidemiology and the Slave Trade," *Political Science Quarterly* 83 no. 2 (June 1968):203–11; *Image of Africa*, p. 197; and "White Man's Grave," pp. 103–10.

9. India Office Records, L/MAR/C 582, pp. 597–600.

10. Jaime Jaramillo-Arango, *The Conquest of Malaria* (London, 1950), pp. 5–12; and Michael Colbourne, *Malaria in Africa* (Ibadan, Nairobi, London, 1966), p. 6.

11. Paul F. Russell, *Man's Mastery of Malaria* (London, 1955), pp. 92–99; Curtin, *Image of Africa*, pp. 192–93 and "White Man's Grave," p. 100.

12. Russell, p. 105; Jaramillo-Arango, p. 87; Colbourne, p. 53.

13. A. Darbon, J.-F. Dulac, and A. Portal, "La pathologie médicale en Algérie pendant la Conquête et la Pacification," pp. 32–38; and Gen. Jaulmes and Lt. Col. Bénitte, "Les grands noms du Service de Santé des Armées en Algérie," pp. 100–103, in *Regards sur la France: Le Service de Santé des Armées en Algérie 1830–1958* (Numéro spécial réservé au Corps Médical, 2ème année, no. 7, Paris, Oct.–Nov. 1958). See also René Brignon, *La contribution de la France à l'étude des maladies coloniales* (Lyon, 1942), pp. 20–21.

14. William Allen, *A Narrative of the Expedition sent by Her Majesty's Government to the River Niger in 1841*, 2 vols. (London, 1848); Paul Merruau, "Une expédition de la Marine Anglaise sur le Niger," in *Revue des Deux Mondes* 1(1849):231–57; Lloyd, p. 150.

15. Curtin, "White Man's Grave," p. 108; Gelfand, *Rivers of Death*, pp. 57–59. Bryson also published an influential article, "On the Prophylactic Influence of Quinine," in the *Medical Times Gazette* of January 7, 1854, cited in Curtin, *Image of Africa*, pp. 355–56.

16. Thomas Joseph Hutchinson, *Narrative of the Niger, Tshadda and Binuë Exploration; Including a Report on the Position and Prospects of Trade up those Rivers, with Remarks on the Malaria and Fevers of Western Africa* (London, 1855, reprinted 1966), pref-

ace and pp. 211–21. See also William Balfour Baikie, *Narrative of an Exploring Voyage up the Rivers Kwóra and Binue (Commonly Known as the Niger and Tsádda) in 1854* (London, 1856, reprinted 1966); Curtin, "White Man's Grave," p. 109; Dike, p. 61 n. 2; Gelfand, *Rivers of Death*, p. 59; Goldsmith, p. 390; Lloyd, pp. 187–98; and Newbury, pp. 73–77.

17. Curtin, *Image of Africa*, p. 362 and "White Man's Grave," pp. 109–10; and Philip Curtin, Steven Feierman, Leonard Thompson and Jan Vansina, *African History* (Boston, 1978), p. 446. In later years, after the discovery of the mosquito transmission of malaria and yellow fever, the death rates fell still further; in the Gold Coast to thirteen to twenty-eight per thousand after 1902, and to less than ten per thousand after 1922.

18. Gelfand, *Livingstone*, passim, and *Rivers of Death*, pp. 63–72; Horace Waller, ed., *The Last Journals of David Livingstone*, 2 vols. (London, 1874), 1:177.

19. On the cinchona transfer to India and Indonesia, see Lucile H. Brockway, *Science and Colonial Expansion: The Role of the British Botanic Gardens* (New York, 1979), pp. 104–33; William H. McNeill, *Plagues and Peoples* (Garden City, N.Y., 1976), pp. 279–80; George Cyril Allen and Audrey G. Donnithorne, *Western Enterprise in Indonesia and Malaya; a Study in Economic Development* (London, 1957), pp. 91–93; Wilfred Hicks Daukes, *The "P. & T." Lands [Owned by the Anglo-Dutch Plantations of Java, Ltd.] An Agricultural Romance of Anglo-Dutch Enterprise* (London, 1943), pp. 43–44 and 105; F. Fokkens, *The Great Cultures of the Isle of Java* (Leiden, 1910), pp. 27–31; and "Cinchona," in *The Standard Cyclopedia of Horticulture*, ed. L. H. Bailey, 3 vols. (New York, 1943), 1:769–71.

20. Newbury, p. 114.

21. On the role of steamers in the British takeover of Nigeria, see A. C. G. Hastings, *The Voyage of the Dayspring. Being the Journal of the Late Sir John Hawley Glover, R.N., G.C.M.G., Together with some Accounts of the Expedition up the Niger River in 1857* (London, 1926); Sir Alan Cuthbert Burns, *History of Nigeria*, 6th ed. (New York, 1963), pp. 95 and 133–34; Goldsmith, p. 391; Dike, pp. 204–12; Lloyd, pp. 128–30 and 199; Newbury, pp. 26, 78, and 114; Davies, pp. 40–48; and D. K. Fieldhouse, *Economics and Empire 1830–1914* (Ithaca, N.Y., 1973), p. 132. On Nigerian resistance, see Obaro Ikime, "Nigeria-Ebrohimi," and D. J. M. Muffett, "Nigeria-Sokoto Caliphate," in Michael Crowder, ed., *West African Resistance: The Military Response to Colonial Occupation* (London,

1971), pp. 205–32 and 268–99; and Robert Smith, "The Canoe in West African History," *Journal of African History* 11 (1970):526–27.

22. Gelfand, *Livingstone*, pp. 126, 165, and 176–81; and Norman Robert Bennett, "David Livingstone: Exploration for Christianity," in Robert I. Rotberg, ed., *Africa and its Explorers: Motives, Methods and Impact* (Cambridge, Mass., 1970), p. 45.

23. Richard Hill, *Egypt in the Sudan 1820–1881* (London, 1959), p. 132.

24. Pierre Gentil, *La conquête du Tchad (1894–1916)*, 2 vols. (Vincennes, 1961), 1:51–63.

25. Lederer, pp. 124–25; Marc Michel, *La Mission Marchand 1895–1899* (Paris and The Hague, 1972).

# PART TWO

# GUNS
# AND CONQUESTS

# CHAPTER FOUR

# Weapons and Colonial Wars of the Early Nineteenth Century

Brethren! Oh! Be not afraid
Heaven your Christian work will aid;
Banish all your doubts and tears,
Rifles cannot fail 'gainst spears.
Take your banner! Onward go!
Christian soldiers, seek your foe,
And the devil to refute,
Do not hesitate to shoot.[1]

Technology is power. It is the power wielded over the natural world, the defense against the hostile elements, the means of using the forces of nature to do one's bidding and improve one's condition. Quinine prophylaxis and river steamers were technologies of this kind.

But technology is also power over people. Those who controlled key technologies—irrigation works in ancient Egypt and Mesopotamia, suits of armor and castles in medieval Europe, for example—wielded great power over their subjects and neighbors. Europeans who entered Africa and Asia in the nineteenth century often did so in the face of hostile populations. The history of imperialism is the history of warfare—of strategy, tactics, and weapons. And in that most popular of historical genres, the history of war, we will find some clues to the causes of the new imperialism.

In the nineteenth century, the armies of the European powers possessed similar weapons, and the outcome of battles was determined by the number of soldiers on each side, or by strategy and tactics. In distant parts of the world, however, European forces faced very different circumstances. There, indigenous armies were often much larger than the invading forces, and knowledge of the terrain favored the local warriors. Furthermore, European colonial armies were limited by economy-minded governments reluctant to commit troops or spend money for military operations that did not noticeably enhance the security of the motherland. Nonetheless, European forces were able to conquer large parts of Asia and Africa—empires of truly Napoleonic proportions—at an astonishingly low cost. What made this possible was the crushing superiority of European firepower that resulted from the firearms revolution of the mid-century.

No period in history produced so dramatic a development of infantry weapons as did the nineteenth century. In terms of effective firepower the disparity between the rifle of World War One and the Napoleonic musket was greater than between the musket and the bow and arrow. Unlike quinine prophylaxis and river steamers, the modern gun was developed almost entirely for use among Europeans and Americans, and its application to colonial warfare was a fortuitous side effect. Yet ironically this new technology changed the balance of power in the non-Western world more than it did in Europe itself.

The development of the modern gun was the product of a complex series of minor advances from many different sources, some of them centuries old. Two stages are of particular importance in this evolutionary process. In the first stage, percussion caps, rifling, oblong bullets, and paper cartridges brought the muzzle-loader to its peak of perfection. The second stage began with the breechloading Prussian needle-gun and culminated in the Maxim gun. The shift from muzzle-loaders to breechloaders in the 1860s was no ordinary technical achievement. It dramatically widened the power-gap between Europeans and non-Western peoples and led directly to the outburst of imperialism at the end of the century. To understand this momentous change, we must consider European and non-Western weapons and tactics and the resulting power-gap both before and after the 1860s.[2]

At the beginning of the nineteenth century the standard weapon of the European infantryman was the muzzle-loading smoothbore musket. It had a flintlock to detonate the powder through a hole in the breech and a bayonet that could be attached to the barrel for hand-to-hand combat. The Brown Bess, which British soldiers used until 1853, was much the same weapon their forefathers had carried at Blenheim in 1704. It had an official range of 200 yards but an effective one of 80, less than that of a good bow. Despite admonitions to withhold their fire until they saw the whites of their enemies' eyes, soldiers commonly shot away their weight in lead for every man they killed. These muskets took at least a minute to load, so to maintain a steady rate of fire on the battlefield, soldiers were drilled in the countermarch, each rank advancing in turn to shoot, then falling back to reload.[3]

One of the most serious drawbacks of the flintlock muskets was their poor firing record. Under the best conditions, they fired only seven out of ten times, and in rain or damp weather they ceased firing altogether. For this reason soldiers were trained to use their weapons as pikes. In 1807, Alexander Forsyth, a Scottish clergyman and amateur chemist, offered a solution to this problem; using the violent explosive potassium chlorate as a detonating powder and a percussion lock instead of a flintlock, he made a gun that could fire in any weather. Tests showed that a percussion lock musket misfired only 4.5 times per thousand rounds, compared to 411 times for a flintlock. After 1814, Joshua Shaw of Philadelphia improved upon Forsyth's invention by putting the detonating powder into little metal caps, thereby simplifying the loading process and making the weapons even more impervious to the elements.[4]

There were those, of course, who deplored the march of progress; as one correspondent wrote to the *Gentleman's Magazine* in 1817:

Sir:

You will forgive my importunity if I take this occasion to add my views on *Mr. Forsyth's Patent Detonating Lock* to those of other recent correspondents. I cannot deny that Mr. Forsyth's invention offers many vulgar advantages, among which the most important are that the gun is made to shoot harder by consequence of the forceful kindling of the powder, and the absence of a touch-hole. Furthermore it will doubtless fire in the most inclement weather. True sportsmen, however, do not require the new lock, for a good flint-lock will answer every conceivable purpose a gentleman might wish. To those who say that it shoots harder, I say, the patent breech flint-lock shoots hard enough; to those who say it shoots faster, I say, if your flint-lock is good, and you have learned to use it, the difference is too trifling to merit attention by true sportsmen; to those who say it fires in violent wind and rain, I say, gentlemen do not go sporting in such weather.

If, moreover, this new system were applied to the military, war would shortly become so frightful as to exceed all bounds of imagination, and future wars would threaten, within a few years, to destroy not only armies, but civilization itself. It is to be hoped, therefore, that many men of conscience, and with a reflective turn, will militate most vehemently for the suppression of this new invention.

I am, Sir, yours &c., &c.,
*An English Gentleman*[5]

Yet, from the military standpoint, the advantages of percussion were evident enough, as this report from the Opium War attests:

A company of sepoys, armed with flintlock muskets, which would not go off in a heavy rain, were surrounded by some thousand Chinese, and were in eminent peril, when two companies of marines, armed with percussion-cap muskets, were ordered up, and soon dispersed the enemy with great loss.[6]

The French army, which in 1822 became the first military force to adopt percussion locks, did not proceed to a massive reconversion until the war scare of 1840. The Woolwich Board approved the first British percussion gun, the Brunswick rifle, in 1836, and three years later the British army began converting its old flintlocks to the new system. In the Opium War, however, most British troops still carried flintlocks, and the army was purchasing flintlocks as late as 1842. Evidently, European armies were commanded by sporting gentlemen in those days.[7]

If percussion made guns fire more consistently, rifling made them fire more accurately. Spiraling grooves inside the barrel made the bullet spin on an axis parallel to the barrel instead of tumbling randomly; this improved the range of guns considerably. The principle of rifling, known since the sixteenth century, was used mostly in hunting and target guns. A muzzle-loading rifle of the early nineteenth century, such as the Pennsylvania-Kentucky rifle of the American frontiersman, had a useful range of 300 yards, about four times that of a smooth-bore musket.

During the War of Independence many American soldiers were armed with hunting rifles. There were also rifle corps in the armies of the French Revolution, and the British created their first Rifle Brigade in 1800. Yet early nineteenth-century rifles were unsuited for the mass warfare of the times. They took four times as long to load as muskets, and they fouled quickly, thereby making them even harder to load. Sportsmen could afford the time and care that these guns required, but ordinary soldiers could not be expected to display such skills in the heat of battle. Napoleon therefore banned rifles from his armies in 1805, calling them "the worst weapon that could be got into the hands of a soldier."

After the Napoleonic Wars, arms experts were once again tempted by the rifle's advantages and tried to overcome its drawbacks. Their goal was to develop a gun as accurate as a rifle and as quick to load as a musket. The solution they found lay not in the gun but in the bullet. An ideal musket bullet should be small enough to slip easily down the barrel when it is loaded yet large enough to grip the rifling snugly on its way out. This would enable it to get the proper spin. To achieve this, the bullet would have to swell at the moment of explosion. In 1823, Captain Norton invented a bullet with a hollow base, and in 1836 the gunmaker William Greener created one with a wooden base plug that would force the bullet to expand; neither worked satisfactorily, however.

Meanwhile the French were experimenting with long bullets that weighed more than spherical bullets of the same caliber. Long bullets could be used only in rifles, for if they did not spin, they would tumble end over end and fly erratically. Finally, in 1848, a French army captain, Claude Etienne Minié, combined the two innovations, the hollow base and the oblong shape, into one bullet. His cylindro-ogival bullet proved to be remarkably accurate. At 100 yards a Minié rifle hit the target 94.5 percent of the time, compared with a 74.5 percent rate for the Brunswick; at 400 yards the percentages were 52.5 and 4.5 percent, respectively. In 1849 the French army began issuing Minié rifles to its troops. In 1853 the British began replacing their Brown Bess with the Enfield, a new rifle with an official range of 1,200 yards and an effective one of 500, six times that of its precursor. It also used paper cartridges containing the bullet and the correct amount of gunpowder; tallow was used to protect the cartridge from moisture.

Since Europe was at peace when the Minié and Enfield appeared, these guns were tested in the colonies. The French issued their new rifles to the elite Chasseurs d'Afrique who fought in Algeria. The British first tried out the "minny ball" against the Xhosa in the Kaffir War of 1851–52. The Enfield gained its most lasting notoriety in India, however. The revulsion of the Indian sepoys at using a cartridge greased with animal fat sparked the Indian revolution of 1857.[8]

Before turning to the gun revolution of the late nineteenth century, let us pause to consider the global balance of firepower before that time. Much has been said in the histories of imperialism about the slow pace of European expansion in the first half of the century, a slowness that has been attributed variously to negligence, free trade, conservatism, and other forms of lack of motive. By considering the warfare of this period and contrasting it with that of the later decades of the century, we shall see that this phenomenon was also closely linked to the evolution of weapons and to the growing disparity of power between the West and the non-West in the age of the new imperialism.

The British conquest of India was very long and, for the Indian people, very costly. Part of the reason can be attributed to the vastness of the subcontinent. But the changing balance of power between the British and the Indian states was equally responsible. Through centuries-old contact with Europeans, the states of India had learned to raise armies on the European model, which were often trained by European instructors. In a recent article analyzing the Anglo-Indian wars of the eighteenth and early nineteenth centuries, Gayl Ness and William Stahl have shown the remarkable evolution of this balance of power.[9]

In the Mysore Wars at the end of the eighteenth century, British forces of 10,000 to 15,000 easily defeated Indian armies six or seven times as large. The British advantage did not come from their weapons, for the Indian forces had equally good muskets, cannons, and ammunition. Rather, it lay in the modern bureaucratic organization of the British armies in the

face of the armies of the Indian states, "aggregates of individ-
ual heroic warriors" tied together only by personal loyalties.
In the Mahratta Wars of the early nineteenth century, the
British armies defeated forces only twice as numerous. Finally,
to defeat their enemies in the Sikh Wars of the 1840s, the
British had to bring into action armies equal in size and superior in artillery firepower. What had happened was that the
Mahrattas and, to an even greater extent, the Sikhs had begun
to copy the European models of organization, recruitment,
taxation, and government, and the resistance they offered the
British had increased proportionally.

China, as Joseph Needham and his colleagues have demonstrated, led the world in most fields of technology until the
fifteenth century.[10] Yet when China confronted the West in
the nineteenth century, it was with weapons outdated by one
or two centuries. The great forts at Taku, the Bogue, and
other points along the coast were her main defense against the
sea barbarians. The better forts had plain walls without ditches,
bastions, or embrasures. Others were made of sandbags, mud,
and overturned boats. Their cannons were very heavy, made
for cannonballs up to thirty-seven pounds, but they were quite
old; some had been cast for the Ming emperors, three centuries
earlier, by Jesuit priests. Worst of all, they were fixed in the
masonry and could not be aimed at a moving target. In place
of field artillery, the Chinese used gingals, large muskets that
fired iron scrap or balls of up to one pound; these devices
made a deafening roar but did not carry far. As for infantry
soldiers, most of them carried bows, crossbows, swords, spears,
pikes, or even stones, relying on shields to protect them. Only
a few carried guns, poorly made matchlocks that they set on
a tripod to fire, and that they reloaded without wadding or
ramming the bullet. Finally, their war-junks, shallow but hard
to maneuver, were armed with small cannons firing balls up to
ten pounds; these guns were attached to blocks of wood and,
like the fortress guns, could not be aimed. Other naval weap-

ons included boarding nets, stink-pots of burning pitch hurled by hand, and burning fire-boats set adrift to run into enemy ships.[11]

The Burmese, like their Chinese neighbors, also fell considerably short of the military technology exhibited by their European enemies. Their stockades were generally of wood and bamboo, and had few cannons. Their foot-soldiers were ill-equipped as well. Some carried large two-handed swords or spears. Others had matchlocks or flintlocks but were expected to make their own gunpowder. Their bullets, of hammered iron, fit loosely in the barrel and had a short range, as Captain Frederick Marryat explained: "When these muskets do go off (and it is ten to one they do not), it is again ten to one that the bullet falls short, from the inefficacy of the powder." And he concluded:

> . . . if the Burmahs had been as well provided with every species of arms equal to our own, the country would not have been so soon subjugated as it was. Their system of defence was good, their bravery undoubted, but they had no effective weapons.[12]

Algeria provides another example. Though much smaller and less valuable than India, this colony cost France far more to conquer than India cost Britain. The peoples of Algeria had a long, mostly hostile association with Europeans. They had guns as good as the Europeans' and were equally proficient at using them. The French expeditionary force that attacked the city of Algiers in 1830 was heavily equipped in the Napoleonic manner. The infantry carried muskets, the cavalry had lances and swords, and the artillery was particularly powerful, with thirty twenty-four-pounders, twenty sixteen-pounders, and dozens of smaller cannons, howitzers, mortars, and rockets. The biggest novelty, and the only one that would have surprised Napoleon, was the steamer *Sphinx,* which helped bombard the city.[13] The Turkish janissaries and Algerian militiamen defending Algiers had equally good flint lock muskets, but their artillery was antiquated, and their defenses soon collapsed under the French bombardment.[14]

The resistance offered by the Algerians stiffened as the French moved inland. The most important chief of the interior, the emir Abd-el Kader, was determined to oust the invaders. At first the French tried to buy his friendship, and in 1833, General Desmichels gave him 400 new muskets and ammunition. A treaty signed the next year between Desmichels and Abd-el Kader guaranteed the Algerian the right to purchase arms, powder, and sulphur. In June 1835, Abd-el Kader's army, now well armed, attacked and defeated a French force of 2,500. Again, to obtain peace, the French had to offer weapons. By the secret protocol to the Treaty of Tafna in June 1837, General Bugeaud promised the emir 3,000 rifles and ammunition. Thus, to secure their small enclaves, the French provided Abd-el Kader with guns equal to their own. By 1840 the French were confined to the coastal cities, while Abd-el Kader, with an army of 80,000, controlled the interior. He also purchased some 2,000 rifles from Britain, which were smuggled in via Gibraltar and Morocco. With the help of French and Spanish artisans, he began manufacturing his own rifles at Tagdempt and cannons at Tlemcen as well. When Bugeaud (now a marshal) decided to take the offensive after 1840, the French persuaded Britain to cut off arms supplies to Abd-el Kader. Even then the conquest required over 106,000 men, or one third of the French army. By the time Algeria was completely pacified in 1857 after many massacres and scorched-earth compaigns, France had lost 23,787 men in action, and thousands more from disease.[15]

What conclusions can we draw from these four examples of early nineteenth-century imperialist warfare? In the First Anglo-Burmese and Opium wars, the Burmese and Chinese had two weaknesses; they were armed with antiquated weapons and—as we saw earlier—they were vulnerable to attack by river steamers. Thus the British were able to achieve their objectives with modest forces in a relatively brief time. India and Algeria were quite different. In the early nineteenth century the Indian states and Algerian tribes possessed infantry weapons comparable to those of the Europeans. If at first they did not have the most efficient organization, they soon developed it. And both were cases of warfare in areas where river steamers were not essential. Consequently, the European forces were eventually reduced to fighting on an equal footing in the midst of enemy territory. For this reason the conquests of India and Algeria were lengthy, costly, and difficult. In contrast to Burma and China, India and Algeria may serve as examples of imperialist warfare without the benefit of technological supremacy. The motivation of the Europeans was there, as was their willingness to sacrifice lives and treasure. What was lacking was the advantage that steamers gave the Europeans in Burma and China, or that the breechloader was to give them later, in Africa.

## NOTES

1. Satirical "hymn," in *Truth* (Apr. 16, 1891) quoted in John Galbraith, *Mackinnon and East Africa 1878–1895; a Study in the 'New Imperialism'* (Cambridge, 1972), p. 15.
2. Sources consulted on early nineteenth-century European firearms include William Young Carman, *A History of Firearms from Earliest Times to 1914* (London, 1955); Russell I. Fries, "British Response to the American System: The Case of the Small-Arms Industry after 1850," in *Technology and Culture* 16(July 1975):377–403; Brig. J. F. C. Fuller, *Armament and History* (New York, 1933); William Wellington Greener, *The Gun and Its Development; with Notes on Shooting*, 9th ed. (London, 1910); Robert Held, with Nancy Jenkins, *The Age of Firearms, a Pictorial History*, 2nd ed. (New York, 1978); James E. Hicks, *Notes on French Ordnance, 1717 to 1936* (Mt. Vernon, N.Y., 1938); Edward L. Katzenbach, Jr., "The Mechanization of War, 1880–1919," in Melvin Kranzberg and Carroll

W. Pursell, Jr., eds., *Technology in Western Civilization*, 2 vols. (New York, 1967), 2:548–51; J. Margerand, *Armement et équipement de l'infanterie française du XVIe au XXe siècle* (Paris, 1945); Col. Jean Martin, *Armes à feu de l'Armée française: 1860 à 1940, historique des évolutions précédentes, comparaison avec les armes étrangères* (Paris, 1974); H. Ommundsen and Ernest H. Robinson, *Rifles and Ammunition* (London, 1915); Thomas A. Palmer, "Military Technology," in Kranzberg and Pursell, 1:489–502; "Small Arms, Military," in *Encyclopaedia Britannica* (Chicago, 1973), 20: 665–78; and G. W. P. Swenson, *Pictorial History of the Rifle* (New York, 1972).

3. Greener, p. 624; Swenson, p. 12.

4. Swenson, p. 19; Held, pp. 171–74; Greener, pp. 112 and 117; "Small Arms," 20:668; Fuller, p. 110; and Martin, pp. 58–64.

5. Quoted in Held, p. 173.

6. Lieut. Gen. Lord Viscount Gough to *London Gazette* (Oct. 8, 1841) quoted in Fuller, p. 128 no. 20.

7. Greener, p. 624; Carman, p. 178; Margerand, p. 114; Martin, pp. 64–70; Hicks, p. 21; Held, p. 182; Peter Ward Fay, *The Opium War, 1840–1842; Barbarians in the Celestial Empire in the Early Part of the Nineteenth Century and the War by Which They Forced Her Gates Ajar* (Chapel Hill, N.C., 1975), p. 130; John D. Goodman, "The Birmingham Gun Trade," in Samuel Timmins, ed., *The Resources, Products, and Industrial History of Birmingham and the Midland Hardware District* (London, 1866), pp. 384–85.

8. On the development of the rifle and the oblong bullet in the early nineteenth century, see Goodman, p. 385; Carman, pp. 104–13; Fuller, pp. 110 and 128–29 n. 23; Greener, pp. 623–31 and 727; Held, p. 183; Hicks, pp. 31–32; Margerand, p. 116; Martin, pp. 73–82; Ommundsen and Robinson, pp. 18–22, 46–65, and 78–79; "Small Arms," 20:669; and Swenson, pp. 16–25. Examples of the Brown Bess, Minié, Enfield, and other early nineteenth-century military guns can be seen in the Salle Louvois, Musée de l'Armée, Hôtel des Invalides, Paris.

9. Gayl D. Ness and William Stahl, "Western Imperialist Armies in Asia," in *Comparative Studies in Society and History* 19 no. 1 (Jan. 1977):2–29. See also E. R. Crawford, "The Sikh Wars, 1845–9," in Brian Bond, ed., *Victorian Military Campaign* (London, 1967), pp. 35–36.

10. Joseph Needham et al., *Science and Civilisation in China*, 7 vols. in 12 parts (Cambridge, 1954–).

11. On Chinese weapons at the time of the Opium War, see Jack Beeching, *The Chinese Opium Wars* (New York 1975), pp. 51–52; Fay, pp. 209, 272–73 and 344–45; Greener, pp. 123–24; John Lang Rawlinson, *China's Struggle for Naval Development, 1839–1895* (Cambridge, Mass., 1967), pp. 6–15; G. R. G. Worcester, "The Chinese War Junk," *Mariner's Mirror* 34(1948):22; and Barton C. Hacker, "The Weapons of the West: Military Technology and Modernization in 19th-Century China and Japan," in *Technology and Culture* 18 no. 1 (Jan. 1977):43–47. On display in the Cour d'Honneur, Hôtel des Invalides, Paris, are two Chinese bronze smoothbore twenty-four-pounder cannons taken at the Taku forts in 1858; they are very similar to European cannons of the eighteenth century.

12. Capt. Frederick Marryat, *Olla Podrida* (Paris, 1841), pp. 80–82. See also Maung Htin Aung, *A History of Burma* (New York and London, 1967), p. 213. At the end of their article on "Western Imperialist Armies in Asia," Ness and Stahl glance briefly at the three Anglo-Burmese Wars (pp. 22–23); by mentioning neither Burmese weapons nor such British weapons as river steamers, breechloaders, or machine guns, they mistakenly apply the Indian experience to Burma; their conclusion that "technology alone does not appear as a critical variable" is, in this case, in error.

13. Duncan Haws, *Ships and the Sea: A Chronological Review* (New York, 1975), p. 126. There are two models of this ship in Paris: one in the Musée de la Marine (along with a model of its engine), the other in the Musée de la Technique, Conservatoire National des Arts et Métiers.

14. One of the cannons of Fort l'Empereur in Algiers was taken by the French in 1830 and is now on display in the Cour d'Honneur, Hôtel des Invalides, Paris. It is a bronze smoothbore twenty-four-pounder cast in Algiers by the Ottomans in 1581.

15. On the weapons of the French conquest of Algeria, see Pierre Boyer, *La vie quotidienne à Alger à la veille de l'intervention française* (Paris, 1963), p. 140; Raphael Danziger, *Abd al-Qadir and the Algerians: Resistance to the French and Internal Consolidation* (New York, 1977), pp. 25, 117, and 224–56; Charles-André Julien, *Histoire de l'Algérie contemporaine,* Vol. 1: *La Conquête et les débuts de la colonisation (1827–1871)* (Paris, 1964), pp. 53, 178–82, and 279; and Maj. George Benton Laurie, *The French Conquest of Algeria* (London, 1909), pp. 21–36, 91–101, and 208–09.

# CHAPTER FIVE

# The Breechloader Revolution

The military rifles of the 1850s were far superior to their precursors the muskets in both accuracy and consistency of fire. Yet they were slow and awkward to use. They fouled easily, they emitted puffs of smoke, and their paper cartridges were delicate and vulnerable to moisture. Worst of all, they could be loaded only when the soldier was standing, in other words in full view of the enemy. In a battle against less well-armed adversaries the advantages they offered were substantial but not overwhelming. The overwhelming firepower of European colonial armies resulted from another innovation: breechloading.

The idea of loading a gun by the breech, like every other innovation in gun design, occurred repeatedly over several centuries before finally coming to fruition. Some of the earliest working breechloaders—the Ferguson, the Hall, Sharp's carbine—were American guns, the outcome perhaps of the insatiable appetite for firearms in a frontier environment.[1]

In Europe the breechloader gained acceptance very slowly. The ancestor of all European breechloaders was the Dreyse needle-gun. Invented by Johann Nikolaus von Dreyse in the late 1820s, this gun was adopted by the Prussian army in 1841–42 but it was not until 1848 that Prussia replaced her muskets with Dreyses. The gun had a bolt-action breech mechanism and used paper cartridges; a long needle penetrated the cartridge and struck a percussion cap at the base of the bullet, igniting the powder.

The Dreyse's forte was its loading speed. During the war of 1866 between Prussia and Austria over the mastery of the German states, Prussian soldiers, kneeling or lying down, could fire their Dreyses seven times in the span it took the Austrians to load and fire once, standing up. The war, which was brief, culminated in a Prussian victory at Sadowa. This battle not only assured Prussia's supremacy in Germany but revolutionized the art of warfare. Before that battle, armies saw no need to uniformly adopt new equipment; instead they gradually acquired new weapons as their old ones wore out. After Sadowa, however, all the great powers of Europe scrambled to switch to breechloaders. An arms race had begun, renewed every few years by the now rapidly evolving technology of gun design. Armies called for newer and better guns, and their own laboratories competed with private inventors to meet their needs. Long before the generals realized the implications of the new firepower, the fate of nations in war came to depend on the arms manufactured during times of peace. As for the other peoples of the world, their fate now depended partly on the increasing firepower of the great nations of Europe and partly on the waves of discarded obsolete weapons that flooded the international arms market every few years.

In 1866, the year of Sadowa, the French army adopted the Chassepot, a bolt-action needle-gun that fired six times a minute and had an official range of 650 yards, compared to the Dreyse's 350. Old but still serviceable Minié muzzle-loaders were converted to breechloaders as well. Both types of rifles were used in the Franco-Prussian War but neither satisfied the experts. They both fouled quickly and leaked hot gases at the breech, and the more they fouled the more they leaked, until soldiers had to fire them at arm's length to avoid being singed. This annulled any accuracy the rifles might have possessed.

It was the British who solved this problem. Rather than throw out muzzle-loading Enfields of recent vintage, the British army parsimoniously converted them to breechloaders by adding a breech-block mechanism invented by Jacob Snider of New York. Colonel Boxer, superintendent of the Royal Laboratory at Woolwich Arsenal, realized that most of the disadvantages of breechloaders came from the cartridge. In 1866 he developed the brass cartridge, which not only protected the powder in transit, but completely sealed the breech at the moment of firing. The bullet could now be made harder and tighter and the grooves shallower than before, without fear of gas leakage. As a result the bullet had a longer and flatter trajectory; the Snider-Enfield had a range three times as great as that of the Dreyse. In 1869 the British army abandoned the idea of converting its Enfields and instead adopted a brand-new model, the Martini-Henry. This was the first really satisfactory rifle of the new generation: fast, accurate, tough, impervious to the weather, a weapon that made every other gun obsolete. Of course the French and the Prussians did not lag far behind. As soon as their war was over, they too began to search for a better weapon. The French army was first to rearm, adopting the Gras in 1874. Three years later the Prussians followed with the 1877 Mauser. Soon even smaller European countries followed suit.[2]

No sooner had the great powers adopted the latest breechloaders than they had to rearm again, this time with repeating rifles. The earliest repeaters were American: the Smith and Wesson of 1855, the Spencer and the Henry of the Civil War, and the 1867 Winchester. Though quick firing, they had an unfortunate tendency to explode if one bullet touched another in a certain way. In European armies, several systems were designed to contain the bullets and feed them safely to the breech. The French army began converting its Gras single-shot rifles into Gras-Kropatchek repeaters, using a seven-cartridge tubular magazine under the barrel. The Germans followed with various Mausers of the 1880s, and the British offered the Lee-Metford, which used James Lee's patent box-cartridge of 1879.

Just as important as the repeating mechanisms was the invention of smokeless powder. In 1885 the Frenchman Vieille discovered the explosive properties of nitrocellulose. This chemical, and its relative nitroglycerine, burned without smoke or ash, thus enabling soldiers to remain invisible and reducing considerably the need to clean the barrel. Smokeless powder had other advantages as well. It burned more evenly, contained more energy than gunpowder, and was less vulnerable to moisture; for use in colonial wars, the British army even developed a specially stable explosive called Cordite.

Throughout the century, the caliber of guns (that is, the inside diameter of the barrel) diminished as their accuracy improved: from .75" (19 mm) for the Brown Bess, to .584" (14.6 mm) for the Enfield, to .44–.46" (11–11.5 mm) for such repeating rifles as the Martini-Henry or the Gras-Kropatchek. The new explosives, thanks to their higher muzzle velocities, allowed gunmakers to reduce the caliber still further, to .32" (8 mm) or even less for the repeaters of the 1890s, while increasing their range and accuracy.

The fact that European armies could obtain millions of complex new rifles on short order is of course an astonishing phenomenon in itself. Contributing to this feat was the revolution in the making of guns brought on by industrialization. To describe the evolution of the gun industry would take us too far afield. Suffice it to say that its two main components were the "American system" and steel. The "American system" of interchangeable parts was first applied to gun making by Eli Whitney in 1797 but did not reach the Old World until some sixty years later. Much impressed with the American guns shown at the Great Exhibition of 1851, officials of the Woolwich Arsenal sent a delegation to observe the Springfield Armory in Massachusetts. The new Enfield Arsenal was built to make Enfields with interchangeable parts by mass production methods. As for steel, it was an economic factor that made it the choice material for military guns. Before 1875, rifle barrels were commonly made of wrought iron or, in high-quality sporting guns, or mixed iron and steel. With the introduction of the new Bessemer, Siemens-Martin, and Gilchrist-Thomas steel-making processes in the third quarter of the century, the cost of crude steel dropped by three quarters or more, and it became possible to produce high-quality barrels cheaply enough for military use. These factors (as we will see) were of the greatest importance in Asia and Africa, for it meant that the newer European guns could no longer be copied or repaired by blacksmiths but demanded industrial steel and machine shops.[3]

No description of the gun revolution can be complete without mentioning the machine gun. The first machine guns were clumsy hand-cranked, multibarreled devices such as the Gatling of the American Civil War, the Montigny Mitrailleuse which failed to save France in 1870, and the Hotchkiss, Gardner, and Nordenfelt of the late 1870s. These guns were heavy and prone to failure, and served best on boats. Only with the invention of smokeless explosives could a reliable machine gun emerge. This was the brainchild of Hiram S. Maxim. The gun he patented in 1884 used only one barrel. As smokeless cartridges were consistent in their energy output, the loading could be done by gas pressure or by recoil, instead of with a crank. The Maxim was light enough for infantry to carry, it could be set up inconspicuously, and it spat out eleven bullets per second. Other manufacturers soon followed suit with small-caliber, smokeless, quick-firing machine guns of their own. The 1892 Nordenfelt, made at the Enfield Armory, had five barrels and fired ten rounds per second, a rate three times faster than that of its gunpowder precursor.

The gun revolution that had begun in the 1860s was completed by the 1890s. Any European infantryman could now fire lying down, undetected, in any weather, fifteen rounds of ammunition in as many seconds at targets up to half a mile away. Machine gunners had even greater power. Though the generals were not to realize it for many decades, the age of raw courage and cold steel had ended, and the era of arms races and industrial slaughter had begun.[4]

Colonial troops in Asia and Africa were among the first beneficiaries of the gun revolution. In a few instances they received new weapons even ahead of their counterparts stationed in Europe, and occasionally these colonial forces contributed in small ways to the process of technological change.

Until the mid-1870s the French colonial forces were armed with Chassepots or with an earlier gun, the muzzle-loading Fusil de Marine et Tirailleurs Sénégalais of 1861. In 1874–76 the Infanterie de Marine (France's colonial army outside Algeria) was armed with Gras single-shot breechloaders. The British meanwhile had become interested in machine guns and had purchased a number of .45 caliber Gatlings for their colonial troops; these guns saw service in the wars against the Zulus in 1871 and 1879, and against the Ashanti in 1874. Later, in 1884, the British used Nordenfelts in the campaign of Egypt.

After 1878, repeaters made their appearance in the colonies. The first was the Gras-Kropatchek of 1878, which was used in the French attack on Tonkin in 1884 and in the campaigns in the upper Senegal and upper Niger regions in 1885–86.[5] This rifle was followed by the newer Lebel, which saw action in the Sudan beginning in 1892. The new rifles offered greater advantages to colonial troops than to forces in Europe. Their brass cartridges and smokeless explosives were much more resistant to long-distance transport and tropical climates than were paper cartridges and gunpowder. They also weighed one third less and thus required one third fewer porters to carry them through the bush.

The 1890s saw the appearance on colonial battlefields of Maxim and other light machine guns. Lord Wolseley, conqueror of the Ashanti, paid Hiram Maxim a visit in 1885 and "exhibited the most lively interest in the gun and its innovator; and, thinking of the practical purposes to which the gun might be put, especially in colonial warfare, made several suggestions to Mr. Maxim." The German General Staff, still strongly opposed to machine guns in the German army, allowed a few to be sent to Africa to be used by colonial troops.[6] The machine gun was to prove as decisive in the colonial wars of the turn of the century as the breechloader had been in the seventies and eighties.

By the turn of the century there was some experimentation in weapons design for colonial use. The French army, for example, developed its "Modèle 1902," an improved Lebel, especially for the Tirailleurs Indochinois; this gun later evolved into the 1907 model "Fusil Colonial" issued to all colonial troops, which in turn became the basic weapon of French infantrymen in World War One. In addition to these infantry weapons, colonial units possessed a few light howitzers and cannons with explosive shells to knock down the walls of fortified towns; these pieces replaced the Congreve rockets used earlier in the century.[7]

The last bit of "progress" in the evolution of the gun arose in response to the special needs of empire. In the words of the historians Ommundsen and Robinson:

> . . . savage tribes, with whom we were always conducting wars, refused to be sufficiently impressed by the Mark II bullet; in fact, they often ignored it altogether, and, having been hit in four or five places, came on to unpleasantly close quarters.

The solution to this unpleasantness was patented in 1897 by Captain Bertie-Clay of the Cartridge and Percussion Cap Factory at Dum-Dum outside Calcutta: the mushrooming or "dum-dum" bullet. This particular device was considered so cruel that even the "civilized" nations refused to use it against one another.[8]

## NOTES

1. The breechloaders of the seventeenth and eighteenth centuries displayed in the Musée de l'Armée in Paris were experimental guns never produced in quantities.

2. On early breechloaders, see William Young Carman, *A History of Firearms from Earliest Times to 1914* (London, 1955), pp. 117–21; J. F. C. Fuller, *Armament and History* (New York, 1933), p. 116; William Wellington Greener, *The Gun and its Development; with Notes on Shooting*, 9th ed. (London, 1910), pp. 119–22, 590, 631, and 701–11; Robert Held, with Nancy Jenkins, *The Age of Firearms, a Pictorial History*, 2nd ed. (New York, 1978), p. 184; James E. Hicks, *Notes on French Ordnance, 1717 to 1936* (Mt. Vernon, N.Y., 1938), p. 26; J. Margerand, *Armement et équipement de l'infanterie française du XVIe au XXe siècle* (Paris, 1945), p. 117; Col. Jean Martin, *Armes à feu de l'Armée française: 1860 à 1940, historique des évolutions précédentes, comparaison avec les armes étrangères* (Paris, 1974), pp. 124–86; H. Ommundsen and Ernest H. Robinson, *Rifles and Ammunition* (London, 1915), pp. 65–90; Thomas A. Palmer, "Military Technology," in Melvin Kranzberg and Carroll W. Pursell, Jr., eds., *Technology in Western Civilization*, 2 vols. (New York, 1967), 1:493–94; "Small Arms, Military," in *Encyclopaedia Britannica* (Chicago, 1973), 20:670–71; and G. W. P. Swenson, *Pictorial History of the Rifle* (New York, 1972), pp. 19–32.

3. On repeaters, see Carman, pp. 112–22 and 178; Russell I. Fries, "British Response to the American System: The Case of the Small-Arms Industry after 1850," in *Technology and Culture* 16(July 1975):386–87; Fuller, pp. 111 and 120; Greener, pp. 241, 283–84, 590, 703–17, and 727–31; Held, p. 184; Hicks, pp. 27–28; Margerand, p. 118; Martin, pp. 247–328; Ommundsen and Robinson, pp. 54–65, 78–101, 111–12, and 118; "Small Arms," 20:671–77; and Swenson, pp. 24–35.

4. On early machine guns, see Lt. Col. Graham Seton Hutchison, *Machine Guns, Their History and Tactical Employment (Being Also*

*a History of the Machine Gun Corps, 1916–1922*) (London, 1938), pp. 31–50; and "Machine Gun," in *Encyclopaedia Britannica* (Chicago, 1973), 14:521–26.

5. Col. Henri-Nicolas Frey, *Campagne dans le Haut Sénégal et dans le Haut Niger (1885–1886)* (Paris, 1888), pp. 60–62.

6. Hutchison, pp. 54–55.

7. Frey, pp. 52–53.

8. Col. Charles E. Callwell, *Small Wars: Their Principles and Practice*, 3rd ed. (London, 1906), pp. 378 and 438–39; Carman, p. 81; John Ellis, *The Social History of the Machine Gun* (New York, 1975), pp. 82 and 98; Hicks, pp. 27–28; Hutchison, pp. 31–39 and 54; Margerand, pp. 117–18; Ommundsen and Robinson, pp. 93 and 118; Yves Person, *Samori: Une Révolution Dyula*, 3 vols. (Dakar, 1968), 2:907.

# CHAPTER SIX

---

# African Arms

The great rush of conquests that historians call the new imperialism took place in Africa during the last quarter of the nineteenth century. Several factors influenced the timing of the scramble for Africa: the discovery of quinine prophylaxis, the Franco-Prussian War, the ambitions of King Leopold of Belgium, and the economic crisis of the 1880s, to name but a few. The techniques and patterns of conquest, however, were also a product of the gun revolution and the resulting disparity of firepower between Africans and Europeans. We must therefore now direct our attention to Africa and the weapons of the African peoples.

The armament of Africans in the nineteenth century varied considerably from one region to another, depending on cultural, economic, and ecological conditions.[1] We can distinguish roughly three zones, keeping in mind that there was substantial overlap and that the boundaries shifted in the course of the century. Areas that had been in close contact with Europeans for a long time—principally those situated along the West African coast—were well supplied with imported muskets and ammunition. In interior regions where horses could survive, albeit with difficulty, as in the savannas of the Sudan which stretched across Africa from east to west, firearms were known, but rare; here cavalrymen carried swords and spears. Finally, in those areas where animal trypanosomiasis was endemic—a zone that covered most of equatorial, eastern, and southern Africa—there were no horses and very few guns; the principal weapons used were the assegai or throwing spear, and the bow and arrow.

The peoples of the West African coastal states had known firearms for at least three centuries. At the time of the slave trade, imported guns had been one of the main commodities for which slaves were exchanged, and in turn these guns were used to capture more slaves. The importation of firearms did not end with the official abolition of the Atlantic slave trade but continued to grow. In 1829, Britain shipped 52,540 guns to West Africa. In 1844, some 80,530 muskets, 596 pistols, and 6 shotguns reached the African shores. By the 1860s the export of guns from Birmingham to Africa ran about 100,000 to 150,000 per year. From 1845 to 1889, between 6.3 and 17.8 percent of all British gun exports made their way to Africa; only India imported as large a share of British guns. In addition, the gunmakers of Liège, Belgium, produced an average of 18,000 muskets a year for Africa, and in some years over 40,000. France, Spain, and other countries that exported little to India, also manufactured muskets for the African market.

The guns that were shipped to Africa were inferior to those exported to other parts of the world. The so-called "Dane guns" of the African trade were the simplest, cheapest, and worst flintlock smoothbore muskets that European ingenuity could create. While the average British musket sold to India in 1844 cost over two pounds sterling, and several thousand shotguns were sold there at an average price of over six pounds, the British muskets destined for the African trade cost less than half a pound apiece, and those made in Liège cost as little as four shillings. In West African terms, a musket in the 1870s was worth the equivalent of $17\frac{1}{2}$ yards of calico or 13 pounds of gunpowder.[2]

Weapons made so cheaply were as dangerous to their users as to their targets. They seldom fired in wet weather, and they frequently burst. Africans were aware of these defects; in 1855 the Igbo boycotted the Lagos palm oil market for five months, partly to protest the defective guns offered for sale there. Yet in some ways Dane guns were well suited to Africa, for they could more easily by repaired by village blacksmiths than higher-quality weapons. The gunpowder available in Africa, whether imported or locally made with imported sulphur, was also of poor quality. Indigenous gunpowder absorbed moisture and fired inconsistently. Imported gunpowder was corned to resist moisture and burn more easily, but it was more expensive. Because of the general inaccuracy of their barrels and the high price of lead, African hunters often used several bullets of polished stone, shotgun style. Only when the Europeans entered Africa with really accurate rifles did Africans feel the need for the same.[3]

In the Sudanic belt between the Sahara and the West African forests, horses were much prized by the aristocracy, and the cavalry was the mainstay of the power of kings. Warriors on horseback wore light armor of chain-mail, quilted cloth, or leather and fought with spears, swords, and shields. Foot-soldiers shot unfletched poisoned arrows at short range or wielded battle axes and assegais. Towns were protected by walls of adobe. Guns were not unknown. Since the fifteenth century, traders and raiding expeditions from North Africa had repeatedly brought firearms into the Sudan. Armies equipped with guns periodically took part in Sudanic wars. Yet the high cost of imported guns and powder, and the inability of Sudanic blacksmiths to produce their own, made firearms a sporadic phenomenon.[4]

The third region of Africa, from a weapons point of view,

lay roughly south of the fourth parallel North. As one went further from the coast into the interior of central and southern Africa, guns got progressively rarer, a consequence of transportation problems and the trading policies of the coastal states. In those regions of the African interior where the tsetse-borne trypanosomiasis made it impossible to keep horses, states were small and loosely defined, and the main weapons were the bow and arrow and the assegai. Here Africans had little or no experience with firearms when they first met Europeans; as one African survivor of such an encounter recalled: "The whites did not seize their enemy as we do by the body, but thundered from afar. . . . Death raged everywhere—like the death vomited forth from the tempest."[5] In other places the few muskets that existed were used for psychological effect; in David Birmingham's words, "although used with minimal skill, the noise of a large, overloaded musket was a major factor in deciding the outcome of village warfare."[6] It was here also that Muslim slave-raiders from the Eastern Sudan and the Indian Ocean coast, gradually moving forward in the course of the nineteenth century, caused such havoc among the non-Muslim peoples of the interior, though the raiders were armed only with muskets.[7]

The basic cause of sub-Saharan Africans' dependence on imported guns was the nature of their iron industry. Iron was made in village furnaces ventilated either by hand-operated bellows or by the natural updraft of the fire. The wheel was practically unknown, and animals, wind, and falling water were seldom used as sources of energy. Hence Africans did not use mechanically operated bellows or air pumps. Because of poor ventilation, their furnaces never got hot enough to produce cast iron. Instead the spongy iron bloom was repeatedly hammered by blacksmiths into wrought iron or steel. The resulting metal was adequate for hand tools but not consistent enough for barrels or precision parts. There was also a problem of productivity, for the cost of African iron was simply too high, and the supply too low, to meet the demand for guns or to compete with cheap firearms from Europe.[8]

The arrival after 1870 of Europeans with quick-firing breechloaders started an arms race in Africa, for it gave Africans a powerful incentive to acquire these new weapons. On a few occasions Africans were able to purchase weapons directly from the manufacturers; for instance, the Khedive Ismaïl of Egypt sent a mission to Europe in 1866–67 to buy breechloaders and steel artillery.[9] Purchases from general traders and gun runners were more common. As the armies of Europe periodically rearmed with ever more powerful weapons, their obsolete models were sold to private entrepreneurs, who disposed of them throughout the world. Thus when Napoleon III dissolved the Garde Nationale after 1866, its 600,000 muskets were sold to merchants who shipped many to Africa. When the French army rearmed with Gras rifles in 1874, its old Chassepots went to gun manufacturers in Liège, who rebuilt them and sold them to traders; by 1890, Chassepots were freely available along the coasts of Africa.[10] In the 1890s the Fon of Dahomey were able to buy large numbers of Chassepots and other guns left over from the Franco-Prussian and American Civil wars: Dreyses, Mausers, Mitrailleuses Montigny, Spencers, Winchesters, and others.[11] According to British consular reports, German, Portuguese, and French officials connived to introduce firearms and munitions of all kinds into British East Africa.[12]

Black South Africans obtained modern weapons from whites living in their midst. Settlers had up-to-date rifles which they sometimes lost or gave away. Mine operators found that they could best recruit black laborers by offering them rifles. Governments, desperate for revenue, benefited from the high taxes and import duties on firearms and encouraged their proliferation. Even missionaries like Livingstone provided their converts with weapons for self-defense. The ability of a number of

South Africans to resist the European intrusion, as in Lesotho or Griqualand West, can be linked to these arms purchases.[13]

Yet the barriers to the purchase of modern weapons were formidable. One barrier was cost. In 1895, in Rabah's capital of Dikwa in the Central Sudan, a Martini-Henry cost one hundred Maria-Theresa dollars, as compared to three to seven for a slave, Rabah's chief export commodity.[14] And when Mohammed bin Abdullah—the so-called "Mad Mullah"—rose up in Somalia in 1898, he had to trade five or six she-camels for every rifle he bought.[15]

The high price of modern rifles reflected not only the costs of manufacture and shipping but also the many political restrictions to which they were subject. Europeans in Africa were worried about Africans obtaining these weapons and tried to prevent this from happening. From the mid-century on, the white settlers of Natal and Cape Colony proposed many schemes to register firearms, to restrict their sale, or even to disarm Africans entirely. In 1854, Britain agreed to stop sales of breechloaders to Africans but to continue them to the Orange Free State.[16] By the 1880s the influx of breechloaders into Africa caused increasing anxiety among Europeans. Evan-Smith, the British consul-general in Zanzibar, wrote in 1888:

> Unless some steps are taken to check this immense import of arms into East Africa the development and pacification of this great continent will have to be carried out in the face of an enormous population, the majority of whom will probably be armed with first-class breech-loading rifles.[17]

To stem the flow of arms, the Brussels Treaty of 1890 prohibited the sale of breechloaders to Africans between the twentieth parallel North and the twenty-second parallel South, but permitted smoothbore muskets to be sold in this zone. The European colonial powers were not satisfied with this measure, for they followed it up with similar acts in 1892 and 1899. On the whole, however, the policy achieved its goal.[18]

The effect of arms restrictions was most noticeable in the

Central Sudanic kingdoms of Sokoto, Bornu, and Wadai. There, as we have seen, firearms were known, but rare and precious. The reason is not difficult to discern. For arms to reach the Central Sudan, they had to pass through several other states, each of which had reasons to restrict the trade. The French conquest of Algeria after 1830 and the Ottoman occupation of Tripoli after 1835 effectively choked off the gun trade from the north. As European traders penetrated up the Niger after 1854, some of their rifles found their way to Yorubaland and Ibadan. The Brussels Treaty of 1890, however, put a stop to the open importation of modern weapons before the advancing frontier of the rifle trade had reached the Central Sudanic states.

The Sanusi, who took over Cyrenaica and some Saharan oases after 1890, partly compensated for this cut-off by providing a new source of arms. The numbers of weapons found in Wadai in the course of the century show the results of the trade. Between 1836 and 1858, Sultan Muhammad Saleh ash-Sharif seems to have accumulated 300 muskets; Sultan Ali (1859 to 1874) had 4,000 muskets; and Sultan Doudmourra (1902 to 1909) had 10,000 guns of which 2,500 were breechloaders. Doudmourra's troops, though, were not trained to use their weapons. Until the very end, firearms in the Central Sudan were so valuable that rulers entrusted their soldiers with them only on the eve of battle, and powder and ammunition were too precious to be wasted on target practice. Despite a centuries-long acquaintance with guns, the Central Sudanic kingdoms were just entering the age of firearms when the Europeans invaded them.[19]

## NOTES

1. Fortunately, a number of articles have appeared recently on the subject of weapons and warfare in nineteenth-century Africa. Among them are a) In Michael Crowder, ed., *West African Resistance: The*

*Military Response to Colonial Occupation* (London, 1971): Michael Crowder, "Preface," pp. xiii–xiv and "Introduction," pp. 1–18; J. K. Fynn, "Ghana—Asante (Ashanti)," pp. 19–52; B. Olatunji Oloruntimehin, "Senegambia—Mahmadou Lamine," pp. 80–110; Yves Person, "Guinea—Samori," pp. 111–43; David Ross, "Dahomey," pp. 144–69; Robert Smith, "Nigeria—Ijebu," pp. 170–204; Obaro Ikime, "Nigeria—Ebrohimi," pp. 205–32; and D. J. M. Muffett, "Nigeria—Sokoto Caliphate," pp. 268–99. b) In the *Journal of African History:* R. W. Beachey, "The Arms Trade in East Africa," 3 no. 3 (1962): 451–67; M. Legassick, "Firearms, Horses and Samorian Army Organization 1870–1898," 7(1966):95–115; Robert Smith, "The Canoe in West African History," 11(1970): 515–33; Gavin White, "Firearms in Africa: An Introduction," 12 (1971):173–84; R. A. Kea, "Firearms and Warfare in the Gold and Slave Coasts from the Sixteenth to the Nineteenth Centuries," pp. 185–213; Humphrey J. Fisher and Virginia Rowland, "Firearms in the Central Sudan," pp. 215–39; Myron J. Echenberg, "Late Nineteenth-Century Military Technology in Upper Volta," pp. 241–54; Shula Marks and Anthony Atmore, "Firearms in Southern Africa: A Survey," pp. 517–30; Anthony Atmore and Peter Sanders, "Sotho Arms and Ammunition in the Nineteenth Century," pp. 535–44; Anthony Atmore, J. M. Chirinje, and S. I. Mudenge, "Firearms in South Central Africa," pp. 545–56; J. J. Guy, "A Note on Firearms in the Zulu Kingdom with Special Reference to the Anglo-Zulu War, 1879," pp. 557–70; Sue Miers, "Notes on the Arms Trade and Government Policy in Southern Africa between 1870 and 1890," pp. 571–77; Joseph P. Smaldone, "Firearms in the Central Sudan: A Revaluation," 13(1972):591–607; and R. A. Caulk, "Firearms and Princely Power in Ethiopia in the Nineteenth Century," pp. 603–30.

2. Kea, pp. 200–201; John D. Goodman, "The Birmingham Gun Trade," in Samuel Timmins, ed., *The Resources, Products, and Industrial History of Birmingham and the Midland Hardware District* (London, 1866), pp. 388, 419, and 426; White, pp. 176–83; Russell I. Fries, "British Response to the American System: The Case of the Small-Arms Industry after 1850," in *Technology and Culture* 16(July 1975):380 and 398; David Birmingham, "The Forest and Savanna of Central Africa," in John E. Flint, ed., *Cambridge History of Africa* vol. 5: *From c. 1790 to c. 1870* (Cambridge, 1976), p. 264.

3. Beachey, pp. 451–52; K. Onwuka Dike, *Trade and Politics in the Niger Delta 1830–1855: An Introduction to the Economic and Political History of Nigeria* (Oxford, 1956), p. 107; Echenberg, pp. 251–52; Fries, pp. 392–93; Jack Goody, *Technology, Tradition and*

*the State in Africa* (London, 1971), p. 52; Jan S. Hogendorn, "Economic Initiative and African Cash Farming: Pre-Colonial Origins and Early Colonial Developments," in Peter Duignan and L. H. Gann, eds., *Colonialism in Africa, 1870–1960,* vol. 4: *The Economics of Colonialism* (Cambridge, 1975), p. 298; Kea, pp. 203–05; Robin Law, "Horses, Firearms and Political Power," in *Past and Present* 72(1976):113–32.

4. Fisher and Rowland, pp. 215–39. See also Echenberg, pp. 245–54; Pierre Gentil, *La conquête du Tchad (1894–1916),* 2 vols. (Vincennes, 1961), 1:55; Goody, pp. 47–58; Muffett, p. 277; and Thomas R. De Gregori, *Technology and the Economic Development of the Tropical African Frontier* (Cleveland and London, 1969), p. 121.

5. Marks and Atmore, p. 519.

6. Birmingham, p. 262.

7. H. A. Gemery and J. S. Hogendorn, "Technological Change, Slavery and the Slave Trade," in Clive Dewey and Antony G. Hopkins, eds., *The Imperial Impact: Studies in the Economic History of India and Africa* (London, 1978), pp. 248–50; and Dennis D. Cordell, "Dar Al-Kuti: A History of the Slave Trade and State Formation on the Islamic Frontier in Northern Equatorial Africa (Central African Republic and Chad) in the Nineteenth and Early Twentieth Centuries" (Ph.D. dissertation, Univ. of Wisconsin, 1977).

8. Walter Cline, *Mining and Metallurgy in Negro Africa* (Menasha, Wis., 1937), esp. ch. 4 and ch. 10; and Goody, pp. 26–29 and 38. Europeans had used water wheels to power bellows and make cast iron since the late fourteenth century, while the Chinese had been making cast iron since the fourth century B.C. and had water-powered bellows from the first century B.C. on; see Joseph Needham, *The Grand Titration: Science and Society in East and West* (London, 1969), pp. 38–39.

9. R. Hill, *Egypt in the Sudan 1820–1882* (London, 1959), p. 109.

10. Yves Person, *Samori: Une Révolution Dyula,* 3 vols. (Dakar, 1968), 2:908 and 991 n. 48.

11. Kea, p. 213; Ross, p. 158.

12. Beachey, pp. 454–67.

13. Atmore, Chirinje, and Mudenge, pp. 546–53; Miers, pp. 571–72; Marks and Atmore, p. 517; Guy, p. 559.

14. Joseph P. Smaldone, "The Firearms Trade in the Central Sudan in the Nineteenth Century," in Daniel F. McCall and Norman R. Bennett, eds., *Aspects of West African Islam* (Boston, 1971), p. 162.

15. Beachey, p. 464. J. J. Vianney gives a much higher and less

believable figure: 300 camels for a rifle in 1900; see his "Mohamed Abdulla Hasan: A Reassessment," in *Somali Chronicle* (Mogadiscio, Nov. 13, 1957), p. 4, cited in Robert Hess, "The 'Mad Mullah' and Northern Somalia," *Journal of African History* 5(1964):420.

16. Atmore and Sanders, p. 539; Marks and Atmore, p. 524.

17. Beachey, p. 453.

18. Beachey, pp. 455–57; Marks and Atmore, p. 528; Miers, p. 577.

19. Smaldone, "The Firearms Trade" and "Firearms in the Central Sudan," pp. 591–607; Fisher and Rowland, pp. 223–30.

# CHAPTER SEVEN

# Arms Gap and
# Colonial Confrontations

Confrontations between Europeans and Africans after 1870 rank among the most lopsided in history. For Africans these encounters brought bewilderment and hopeless struggles, while for Europeans they resembled hunting more than war. Breech-loaders broke down African resistance as decisively as quinine prophylaxis had overcome the barrier of malaria. To be sure, there was a vast range of forms of encounter, from the lone explorer passing through a village to, at the other extreme, a full-scale military campaign. Yet all were marked by the quality of the weapons involved.

All explorers carried rifles. Some, like Livingstone, Cameron, or Barth, used them only for hunting and self-defense. They made their way by befriending the people through whose lands they traveled, delighting them with their shooting prowess; thus Malamine, a Senegalese associate of Savorgnan de Brazza, used his Winchester repeater to become the most famous hunter in the area around Stanley Pool and a friend of the local chiefs.[1]

Such friendly encounters were the exception, however. All too often the European traveler felt obliged to show off the power of his guns. Hauptmann Kling, a German explorer who traveled through West Africa in 1893, demonstrated the force of his machine gun by destroying a wall. He thus gained the respect of his hosts, if not their love.[2] Gustav Rohlfs, a visitor to Bornu, wrote that when the inhabitants of a village "showed an inclination to oppose forcefully our camping there . . . a few blind shots made them see reason."[3]

Of all the explorers, none saw himself as a conquistador quite as much as Henry Morton Stanley. While crossing Africa from east to west on his expedition of 1877–78, he got into a dispute with the inhabitants of Bumbireh, a village on the shore of Lake Victoria. To make his escape he fired his elephant guns at the villagers armed only with spears and bows and arrows. A few months later he returned with 250 men and summarily decimated the settlement. Sir John Kirk, British consul at Zanzibar, called the incident "unparalleled in the annals of African discovery for the reckless use of the power that modern weapons placed in his hands over natives who never before had heard a gun fired."[4]

Later, traveling down the Congo, Stanley encountered people who welcomed him with spears and arrows. Several times his men chased adversaries off with the fire of their Winchesters and Sniders. He would then pursue them, he wrote,

> . . . up to their villages; I skirmish in their streets, drive them pell-mell into the woods beyond, and level their ivory temples; with frantic haste I fire the huts, and end the scene by towing the canoes into mid-stream and setting them adrift.[5]

On his last big journey, the Emin Pasha Relief Expedition of 1886–88, Stanley brought along 510 Remington rifles with 100,000 rounds of ammunition, 50 Winchester repeaters with 50,000 cartridges, and a Maxim gun, a gift of the inventor. He also brought along 27,262 yards of cloth and 3,600 pounds of beads as trade goods, and, for nourishment, forty porter-loads of the choicest provisions from Messrs. Fortnum and Mason of

Picadilly. At no time in history has the distinction between tourists and conquerors been so blurred.[6]

On the heels of the explorers were the military. In the seventies and eighties European statesmen, asserting their arrogant faith in the ability of their armies to overcome all African resistance, drew lines on maps of the continent to indicate where their future conquests would lie. In 1873–74, General Wolseley defeated the Ashanti, one of West Africa's most powerful kingdoms, with a force of 6,500 men armed with rifles, Gatling guns, and 7-pounder field artillery.[7] Similarly, the army of the Senegalese ruler Mahmadou Lamine, with its spears, Dane guns, and poisoned arrows, was defeated by a French force of 1,400 men armed with Gras-Kropatcheks.[8]

In the 1890s the addition of Maxim guns and quick-firing light artillery to the arsenal of the colonialist troops made European-African confrontations even more lopsided, turning battles into massacres or routs. In 1891, near Porto Novo, a French detachment of 300 men, firing 25,000 rounds of ammunition in a 2½ hour battle, defeated the entire Fon army.[9] In 1897 a Royal Niger Co. force composed of 32 Europeans and 507 African soldiers armed with cannons, Maxim guns, and Snider rifles defeated the 31,000-man army of the Nupe Emirate of Sokoto; though some of the Nupe had breechloaders, their insufficient training caused them to fire over the heads of their enemies.

In Chad a French force of 320, most of whom were Senegalese *tirailleurs,* in 1899 defeated Rabah, reputedly the fiercest of the Sudanese slave-raiders, with his 12,000 men and 2,500 guns.[10] The Caliphate of Sokoto finally fell in 1903 after an attack by a British force of 27 officers, 730 troops, and 400 porters.[11] And in 1908 the 10,000-man army of Wadai was routed by 389 French soldiers.[12]

Perhaps the most famous of all colonial campaigns—at least in the English-speaking world—was General Kitchener's conquest of the Sudan in 1898. The British believed the Sudanese Dervishes were skilled but fanatical warriors and blamed them for having defeated General Gordon in 1885. Kitchener's expedition was therefore well supplied with the latest weapons: breechloading and repeating rifles, Maxim guns, field artillery, and six river gunboats firing high-explosive shells.

At one point the British Camel Corps was almost overwhelmed by a Dervish attack. Winston Churchill, who took part in the campaign, described the British response:

> But at the critical moment the gunboat arrived on the scene and began suddenly to blaze and flame from Maxim guns, quick-firing guns and rifles. The range was short; the effect tremendous. The terrible machine, floating gracefully on the waters—a beautiful white devil—wreathed itself in smoke. The river slopes of the Kerreri Hills, crowded with the advancing thousands, sprang up into clouds of dust and splinters of rock. The charging Dervishes sank down in tangled heaps. The masses in the rear paused, irresolute. It was too hot even for them.

At Omdurman, Kitchener confronted the main Dervish army of 40,000. According to Churchill:

> The infantry fired steadily and stolidly, without hurry or excitement, for the enemy were far away and the officers careful. Besides, the soldiers were interested in the work and took great pains. But presently the mere physical act became tedious. . . .

The Dervish side looked very different:

> And all the time out on the plain on the other side bullets were shearing through flesh, smashing and splintering bone; blood spouted from terrible wounds; valiant men were struggling on through a hell of whistling metal, exploding shells, and spurting dust—suffering, despairing, dying.

After five hours of fighting, 20 Britons, 20 of their Egyptian allies and 11,000 Dervishes lay dead.

> Thus ended the battle of Omdurman—the most signal triumph ever gained by the arms of science over barbarians. Within the space of five hours the strongest and best-armed savage army yet arrayed against a modern European Power had been destroyed

and dispersed, with hardly any difficulty, comparatively small risk, and insignificant loss to the victors.[13]

The general rule in late nineteenth-century Africa, that Europeans easily overcame African resistance through superior firepower, is not without exception. In a few instances—primarily in the Western Sudan and Ethiopia—Africans held back the Europeans for many years. These cases also illustrate the importance of the new weapons.

Samori Touré, whose homeland was the high country between the upper Niger and the upper Senegal, was an upstart who assumed the traditional Islamic role of military-religious leader. He was not hampered by obsolete military customs and, better than any other African ruler of his time, he grasped the importance of modern weapons. He was the first leader in the region to arm all his troops with guns. In the early 1870s he purchased a number of muskets, which gave him a tactical superiority over other Sudanic rulers and helped attract followers. By 1886 he had acquired fifty breechloaders, most of them Chassepots. Begining in 1890, he bought as many Gras and Mauser rifles as he could obtain from dealers in Sierra Leone. After several battles with French troops armed with repeaters—Gras-Kropatcheks and Lebels—Samori made every effort to obtain these weapons by purchase from Sierra Leone or by capture on the battlefield. His supply of repeaters increased from 36 in 1887 to 4,000 in 1898, the year he was finally defeated. During his last ten years he was able to hold out against the advancing French forces through the judicious use of his modern weapons and through the mobility of his army; in fact, his methods came to resemble more closely the guerrilla tactics of late twentieth-century liberation wars than the mass attacks of the Dervishes.

The most interesting aspect of Samori's career was his attempt to create his own arms industry. He sent a blacksmith to work in the arsenal at Saint-Louis and learn French techniques. He employed 300 to 400 blacksmiths who produced 12 guns a week, mostly copies of Gras-Kropatcheks; they also turned out 200 to 300 cartridges a day, while their wives made gunpowder. The manufacture of modern guns demands machine tools and high-grade steel, however, and Samori's homemade weapons were never numerous or precise enough to stem the French advance for very long. In the end he was defeated when the French cut off his access to supplies from Sierra Leone and Libya.[14]

A similar situation prevailed in Ethiopia, but on a larger scale. During the reign of Tewodros (1855–68) various regional armies used muskets, cavalry, and artillery. In 1868 a British force defeated Tewodros's army at Meqdela, thanks in part to their breechloaders. The British returned home, although their technological legacy remained. Tewodros used missionary craftsmen to cast cannons. Bezbiz Kasa, a rival warlord, recruited an English sergeant named Kirkham to train his troops. In 1871, after defeating his rivals, Kasa became Emperor Yohannis IV. The 1880s in Ethiopia were marked by wars between various regional lords and against the Sudanese Mahdists to the west and the Italians to the east. In the course of these wars foreign arms poured into the country. Menelik of Showa was able to obtain breechloaders and ammunition from the Italians at Massawa on the Red Sea. By 1896 he had a better-equipped army than any African ruler had ever had. His arsenal included several thousand breechloaders, a few machine guns, and even some field artillery. When an Italian army of 17,000 men advanced into Ethiopia, it confronted an army equally well equipped and even better trained. The defeat of the Italians at Aduwa on March 1, 1896, resulted as much from their own Treaty of Wichelle, by which they had agreed to supply Menelik with arms, as from any tactical blunders on the battlefield.[15]

Not just weapons but also the strategy and tactics involved in colonial wars deserve our attention for what they reveal both about the armies that confronted one another and about the thinking of European military theorists before World War One. European forces in Asia and Africa broke most of the classic rules of war. In forest and bush country they often had to advance single-file along narrow paths. They were totally

dependent on exposed supply lines stretching for many miles over difficult terrain; sometimes they had to bring along food and water for porters and pack animals as well as for soldiers.[16] Under such circumstances they would have been vulnerable to guerrilla tactics. In a few cases, as in Algeria and on the north-west frontier of India, indigenous resistance forces using guerrilla tactics were able to tie down large numbers of imperialist troops for long periods of time. But guerrilla tactics were rare, because they demanded a more flexible social structure and a higher level of political consciousness than were common in most non-Western societies of the time.

The method of fighting that Europeans most frequently encountered, from the Sudan to South Africa and from West Africa to China, was the frontal assault, or rush, by large numbers of fighting men. These warriors often displayed superior courage, and their tactics and discipline were appropriate to the kinds of warfare they were accustomed to. But against modern rifles these methods were obsolete. Firing on the run, reloading standing up, or dashing up to hurl a javelin at close range were, under these new circumstances, suicidal. Against the open assault of masses of warriors, the imperialist forces resurrected the square of Napoleonic times, a human fortress surrounded by an impenetrable hail of bullets. It was a near-invincible defense against attacking forces armed with inferior weapons, no matter how numerous.[17]

Such a battle took place in October 1893 near Zimbabwe in southern Africa. A column of 50 British South African Police had encountered the 5,000 Ndebele warriors of King Lobengula. The Ndebele carried assegais and shields; the British had four Maxims, a Nordenfelt, and a Gardner. Within an hour and a half 3,000 Ndebele lay dead. Lieutenant Colonel Graham Hutchison, a British writer of the purple-prose school of imperial history, described the confrontation:

> Fierce tribesmen, inflamed with racial fanaticism, armed with the assegais, formed their impis, and in great force went forth to battle; while a thousand war-drums, in wild crescendo, beat their primitive tatoo of vengeance amid the scattered kraals. The B.S.A.P., though hurriedly reinforced by volunteer Rhodesians, were from the outset greatly outnumbered. . . . They stood on the defensive, forming a wagon laager, within which had been concentrated women, children and provisions, and provoked the Matabele to charge. Maxim guns were placed at the angles of the laager: and it is recorded how again and again hordes of Matabele bit the dust far beyond the thrust of the deadly assegai.[18]

Imperialist warfare presents a paradox. While the strategy of the Europeans was primarily offensive—to seek out the enemy in his own territory, to destroy his forces and his government, and to seize his land—their tactics were primarily defensive. This combination of offensive strategy and defensive tactics was the result of the newly increased firepower of the Europeans and the inevitable time-lag on the part of Africans and Asians in developing the tactical response of guerrilla warfare.

There has been little interest in the art and theory of colonial warfare during the period 1871–1914. Most European military writers, then and now, have considered this a time of peace, and consequently paid scant attention to these "small wars."[19] One of the very few studies of nineteenth-century colonial warfare is *Small Wars: Their Principles and Practice* by Colonel Charles Callwell. In this textbook for officers published in 1906, the author recognized the curious juxtaposition of offensive strategy and defensive tactics but did not pursue its implications. He took for granted the European superiority in firepower, hardly commenting on it.[20] Instead he repeatedly stressed the purported moral differences between the opponents: on the one hand were the European and European-led soldiers, full of zeal, resolution, daring, vigor, boldness, courage, and other noble virtues; on the other, the hordes of barbarians, savages, fanatics or, at best, the semicivilized races.

Another such writer was Charles B. Wallis, a former district commissioner in Sierra Leone; his *West African Warfare,* a manual for British officers posted to that region, was a bit more realistic. He too waxed florid about "the weird and treacherous surroundings, the nerve-wracking effects of the climate," and the "cunning savages . . . like the wild animals of his own forests." He too spoke highly of "the stern discipline and enthusiastic *esprit de corps* of the British army" led by "that indispensable factor in the machine of West African warfare—the British officer." But at least, from personal experience, he knew the value of modern rifles and Maxim guns, and in the event of a pitched battle he advocated the square with Maxims at the corners.[21]

The experience of colonial warfare may help explain the disastrous tactics of World War One. For forty years, the only wars that Britain, France, and Germany fought were colonial wars, and these only confirmed the Napoleonic principle that the key to victory was an offensive strategy backed by overwhelming firepower. What the generals of the First World War misunderstood was that their new rifles and machine guns were defensive weapons, and that colonial victories had been achieved by defensive tactics against poorly armed enemies.

The caste and culture of Callwell, Wallis, and Hutchison erected around the colonial forces a false aura of moral and racial superiority, which hid the technological revolution in tactics. The soldier in the trenches of Flanders, with his rifle or machine gun, was indeed as invulnerable to enemy attack as his counterpart had been in the square at Omdurman or in the wagon laager in Ndebeleland. But conversely, the soldier going over the top of a Flanders trench was as vulnerable as any Dervish or Ndebele warrior. The surprise of World War One was that the offensive had become suicidal, that vigor, élan vital, courage, esprit de corps, and all the other presumed virtues of the European fighting man were irrelevant against the hail of bullets from these same rifles and machine guns.

The effect of modern infantry weapons on the battlefields of Europe was quite the opposite of what it had been in Africa. Instead of bringing about the quick and easy success the European powers had become used to, the new firearms made victory impossible.

# NOTES

1. Henri Brunschwig, *French Colonialism 1871–1914: Myths and Realities,* trans. William Glanville Brown (London, 1966), p. 47.

2. Jack Goody, *Technology, Tradition and the State in Africa* (London, 1971), p. 62.

3. Wolfe W. Schmokel, "Gerhard Rohlfs: The Lonely Explorer," in Robert I. Rotberg, ed., *Africa and its Explorers: Motives, Methods and Impact* (Cambridge, Mass., 1970), p. 208.

4. Eric Halladay, "Henry Morton Stanley: The Opening Up of the Congo Basin," in Rotberg, pp. 242–45; Peter Forbath, *The River Congo: The Discovery, Exploration and Exploitation of the World's Most Dramatic River* (New York, 1978), pp. 278–81.

5. Halladay, p. 244. See also Forbath, pp. 296 and 304–08; and Henry Morton Stanley, *Through the Dark Continent, or the Sources of the Nile around the Great Lakes of Equatorial Africa and down the Livingstone River to the Atlantic Ocean,* 2 vols. (New York, 1879), 1:4 and 2:211–12 and 300.

6. Henry Morton Stanley, *In Darkest Africa, or the Quest, Rescue, and Retreat of Emin, Governor of Equatoria,* 2 vols. (New York, 1890), 1:37–39.

7. John Keegan, "The Ashanti Campaign, 1873–4," in Brian Bond, ed., *Victorian Military Campaigns* (London, 1967), p. 186;

J. K. Fynn, "Ghana–Asante (Ashanti)," in Michael Crowder, ed., *West African Resistance: The Military Response to Colonial Occupation* (London, 1971), p. 40.

8. B. Olatunji Oloruntimehin, "Senegambia–Mahmadou Lamine," in Crowder, pp. 93–105; M. Legassick, "Firearms, Horses and Samorian Army Organization 1870–1898," *Journal of African History* 7(1966):102; and Col. Charles E. Callwell, *Small Wars: Their Principles and Practice,* 3rd ed. (London, 1906), p. 378.

9. Callwell, p. 260; David Ross, "Dahomey," in Crowder, p. 158.

10. Pierre Gentil, *La conquête du Tchad (1894–1916),* 2 vols. (Vincennes, 1971), 1:99.

11. A. Adeleye, *Power and Diplomacy in Northern Nigeria 1804–1906: The Sokoto Caliphate and its Enemies* (New York, 1971), pp. 182–83; Obaro Ikime, *The Fall of Nigeria: The British Conquest* (London, 1977), p. 203.

12. Joseph P. Smaldone, "Firearms in the Central Sudan: A Revaluation," *Journal of African History* 13(1972):591–607; and Humphrey J. Fisher and Virginia Rowland, "Firearms in the Central Sudan," *Journal of African History* 12(1971):223–30.

13. Winston Churchill, *The River War: An Account of the Reconquest of the Soudan* (New York, 1933), pp. 274, 279, and 300. See also Lieut. Col. Graham Seton Hutchison, *Machine Guns: Their History and Tactical Employment (Being Also a History of the Machine Gun Corps, 1916–1922),* pp. 67–70; and Callwell, pp. 389 and 439.

14. Yves Person, *Samori: Une Révolution Dyula,* 3 vols. (Dakar, 1968), 2:905–12, and "Guinea–Samori," in Crowder, pp. 122–23; and Legassick, pp. 99–114. On the French side, see Alexander S. Kanya–Forstner, *The Conquest of the Western Sudan: A Study in French Military Imperialism* (Cambridge, 1969), pp. 10–12.

15. R. A. Caulk, "Firearms and Princely Power in Ethiopia in the Nineteenth Century," *Journal of African History* 13(1972):610–26.

16. On the logistics of colonial campaigns, see Callwell, pp. 57–63 and 86–87; J. J. Guy, "A Note on Firearms in the Zulu Kingdom with Special Reference to the Anglo–Zulu War, 1879," *Journal of African History* 12(1971):567–68; and Crowder, p. 11.

17. See D. J. M. Muffett, "Nigeria–Sokoto Caliphate," in Crowder, p. 290; Bond, p. 25; Crowder, p. 9; and Callwell, pp. 30–31.

18. Hutchison, p. 63. See also Edward L. Katzenbach, Jr., "The Mechanization of War, 1880–1919," in Melvin Kranzberg and Carroll W. Pursell, Jr., eds., *Technology in Western Civilization,* 2 vols.

(New York, 1967), p. 551; and John Ellis, *The Social History of the Machine Gun* (New York, 1975), p. 90.

19. For recent examples of this attitude, see Theodore Ropp, *War in the Modern World*, rev. ed. (New York, 1962) and Maj. Gen. J. F. C. Fuller, *War and Western Civilization 1832–1932* (London, 1939).

20. See Callwell, p. 398.

21. Charles B. Wallis, *West African Warfare* (London, 1906).

# PART THREE

# THE COMMUNICATIONS REVOLUTION

# CHAPTER EIGHT

# Steam and the Overland Route to India

> Like the Romans, the British had always laid tremendous stress on communications; and perhaps the genius of British colonial method lay as much in engineering skill as in administration.[1]

Europeans of the fifteenth and sixteenth centuries who left their homes for distant shores had to be self-reliant, able to survive for months or years without contact with Europe. They accomplished this by provisioning themselves largely with local resources. Nineteenth-century Europeans in Asia and Africa possessed far greater resources than did their predecessors. In exchange, though, they were all the more firmly tied to Europe. Rifles and cartridges, steam engines and canned peas, quinine, official stationery, and a thousand other necessities had to be precision-made in factories and delivered halfway around the world to uphold the might and the comfort of Europeans in distant lands. Even more than goods, information was the lifeblood of European imperialism; business deals, administrative reports, news dispatches, and personal messages sustained the colonizers and assured them the support of their own people. The "new" imperialism of the nineteenth century was not new merely because it had been preceded by "old" imperialisms. It was a qualitatively different phenomenon. For the first time in history, colonial metropoles acquired the means to communicate almost instantly with their remotest colonies and to engage in an extensive trade in bulky goods that could never have borne the freight costs in any previous empire. The world was more deeply transformed in the nineteenth century than in any previous millennium, and among the transformations few had results as dazzling as the network of communication and transportation that arose to link Europe with the rest of the world.

Up to the 1830s, when an Englishman corresponded with someone in India, his letter, carried around Africa on an East Indiaman, took five to eight months to reach its destination. Because of the monsoons in the Indian Ocean, he did not receive a response until two years after he had mailed his letter. By the 1850s a message from London went by train across France, by steamer to Alexandria and from Alexandria to Cairo, by camel to Suez, then by steamer to Bombay or Calcutta, where it arrived thirty to forty-five days after leaving London. The answer took an additional thirty to forty-five days, for a round trip total of two to three months.[2] Twenty years later, if a letter still took a month to reach Bombay, a telegram got there in as little as five hours, and the answer could be back the same day. And in 1924 at the British Empire Exhibition, King George V sent himself a cable; it reached him, having circled the globe on all-British lines, eighty seconds later! Small wonder many Britons of the imperial age thought the world empire they had acquired was a not unreasonable reward for the amazing ingenuity of their industry.

Almost as soon as steam proved itself a feasible means of propelling a ship, some men dreamed of steaming across the oceans, while others mocked the idea as preposterous and contrary to nature. As often happens, both sides were right. The promise of steam was obvious: freedom from the fickle winds that made even the best sailing voyage a gamble against time. But the climb to that achievement was long and difficult. Only in the 1830s did steam-powered ocean crossings prove technologically possible, and it was not until the 1850s that such ventures could justify themselves financially.

The first steamers to go to sea were hybrids, sailing ships with auxiliary engines. The *Savannah,* which crossed the Atlantic from America to England in 1819, used her steam engine for less than 4 days in her 27-day crossing. The steamship *Enterprize,* first of her kind to reach India, used steam for 63 of the 113 days she took to sail from Falmouth to Calcutta in 1825. Finally in 1838 two steamers, the *Sirius* and the *Great Western,* crossed the Atlantic on steam alone, and two years later regular steamship service across the Atlantic began, offering the public an alternative at last to sailing ships.[3]

The reason for the long delay—over a century since steam engines had begun pumping water from English mines—is that the low-pressure steam engine was ill suited to compete with the free wind as a source of energy at sea. Before the 1830s steam machinery was large and delicate and frequently in need of adjustment and lubrication. Boilers, fed seawater, had to be shut down every three or four days so that encrusted salt deposits could be scraped off. High seas frequently damaged the paddle-wheels of steamers. So risky were these vessels that Parliament appointed a committee in 1831 to "take into consideration the frequent calamities by Steam Navigation and the best Means of guarding against their Recurrence. . . ."

Worst of all, steam engines were exceedingly fuel-hungry. Near Britain where coal was cheap, or along the tree-lined Mississippi, this might not pose a big problem. But at sea they had to either bring along an enormous supply of coal or be supplied by sailing ships. Distance from the source of coal of course accentuated the problem. The *Great Western's* 440-horsepower engine, for example, burned 650 tons of coal to cross the Atlantic, an average of eight pounds per horsepower per hour. The first Cunard liner, the *Britannia* of 1840, had a total cargo capacity of 865 tons, of which 640 tons were taken up by coal, which she burned at approximately five pounds per horsepower per hour.[4]

The disadvantages of early steamers relegated them to those tasks where their virtues outweighed the prohibitive cost factor. For long-distance transport this occurred along two corridors. One was between Britain and the United States, where traffic was intense and the wealthy were prepared to pay almost any price to get across the ocean quickly. The other was between England and India.

There have been, historically, three routes between Europe and India. One goes across Egypt, down the Red Sea, and across the Arabian Sea; the second leads across Syria to Mesopotamia, then down the Euphrates and the Persian Gulf to the Arabian Sea; and the third circumnavigates Africa. Each epoch has seen one or another take precedence, depending on naval technology and Middle Eastern politics.

The Cape Route around Africa was favored by the East India Company in the early nineteenth century. It was safe, for Britain had swept the Dutch and French fleets from the seas in the Napoleonic Wars. And it involved no transshipments, no complex and delicate relations with Turks, Egyptians, and Arabs along the way. It was, however, very long. Monsoons, blowing toward India half the year and away the other half, made sailing in each direction easy in one season and impossible the next. Hence it took about two years for an East Indiaman to complete a round trip.

Both routes through the Middle East were called "Overland" by the English. The one through Mesopotamia offered easy sailing between Europe and Syria and again from the head of the Persian Gulf to India. But in between lay a part of the Ottoman Empire inhabited by xenophobic Arab tribes and unreliable Turkish administrators, subject to the sometimes uncertain foreign policy of the Sublime Porte itself. The Red Sea Route offered fewer political difficulties but almost insuperable natural obstacles to sailing ships. The Red Sea was notorious for its fickle winds, long calms, and sudden violent storms. Its bottom was dangerously studded with submerged rocks and coral reefs, its coasts were jagged, and coastal inhabitants welcomed the opportunity to plunder wrecked ships.

Only occasionally did a hardy European traveler or a bundle of urgent mail make its way through the labyrinths of the Overland Route; for the most part the Cape Route had become *the* pathway to the East by the early nineteenth century. But all that was to be changed by the advent of steam in Eastern waters.[5]

To the East India Company, an organization that owed its profits to its monopoly on British trade with the East and its fine fleet of East Indiamen, the long voyage around Africa was a fact of life. Yet, with the volume of trade between India and Britain growing rapidly, businessmen in both countries cried out for a shorter route. Between 1790 and 1817, British exports to the East rose from 26,400 tons to 109,400 tons. British cloth exports to India were especially important, rising from 817,000 yards worth £201,182 in 1814 to 51,833,913 yards worth £3,238,248 in 1832. The entrepreneurs who handled this trade were the catalysts of improved communications in the nineteenth century.

In the year 1822 the naval officer James Henry Johnston, finding much enthusiasm in Britain for steam navigation, decided to initiate a steamer service between Calcutta and Suez. He sailed to India to sell his project to the Anglo-Indians of Calcutta. They needed no prodding. Influenced by the performance of the little *Diana* on the Hooghly River, a group of Anglo-Calcuttans founded a "Society for the Encouragement of Steam Navigation between Great Britain and India." This

"Steam Committee" set up a "Steam Fund" of 69,903 rupees, to which the governor-general of India, Lord Amherst, contributed 20,000 rupees, the nawab of Oudh another 2,000, and various businessmen of Calcutta the rest. The money was offered as a prize to anyone whose steamship could make four consecutive voyages between Bengal and England at an average of seventy days per trip. This implied, of course, taking the Cape Route. Johnston returned to England to build a ship that would win the prize.

Johnston found ample support among London capitalists and proceeded to build the first British steamer designed for the high seas. The aptly named *Enterprize* was large for her day at 141 feet in length and 464 tons burden, and had two 60-horsepower Maudslay engines. Her builders confidently expected the ship to reach Calcutta in sixty days, with one refueling stop at Capetown, and sent her out to sea in August 1825 without even a trial run. Unfortunately she was so slow, and her engines so voracious, that she soon ran out of fuel. All in all she took 113 days to reach Calcutta, of which only 63 were under steam. Thus she forfeited the prize and disappointed the steam enthusiasts of Calcutta. The Bengal government bought her for use in the Burma War, and the Committee, in recognition of a gallant try, awarded half the prize to "Steam" Johnston.[6]

The voyage of the *Enterprize* was a failure for Calcutta but not for all of India. Because of the monsoons, a sailing ship coming from Europe can reach the east coast of India faster than it can the west coast. Steamers, however, go by distance, not by the winds; and if the technology of the 1820s could not produce a long-distance steamer for the Cape Route, it did promise to open up the Overland Routes from Bombay and make that city the new gateway to India. For this reason the Anglo-Indians of Bombay took up the idea of the Overland Route, to the great mortification of the Anglo-Indians of Calcutta. Each community got the support of its government: on the one side, the Presidency of Bombay and the Indian Navy (known until 1830 as the Bombay Marine); on the other, the Presidencies of Calcutta and Madras and the governor-general of India. And wavering between the rivals were the distant East India Company and the British government. Steamship technology, too costly for private enterprise alone, thus became entangled in the labyrinthine politics of British India.[7]

The credit for initiating steam communication between Britain and India must go to the government of the Presidency of Bombay rather than to business entrepreneurs. In 1823 and again in 1825–26, Mountstuart Elphinstone, governor of Bombay, asked the Court of Directors in London to authorize a Red Sea steam service. Receiving no answer, he went ahead with his plans and ordered the Bombay Marine to begin surveying the coasts of the Arabian and Red seas. His successor after 1827, Sir John Malcolm, carried out Elphinstone's plans, ably assisted by two brothers, Sir Charles Malcolm, superintendent of the Indian Navy, and Admiral Sir Pulteney Malcolm, commander of the Royal Navy's Mediterranean squadron. In 1828 coal stocks were deposited along the way to Suez, and the Bombay Marine acquired the *Enterprize*. When that vessel broke down, the Malcolms had a new ship built at Bombay. In mock deference to the Court of Directors, which had vehemently forbidden the whole scheme, they named the ship after its chairman, Hugh Lindsay.

The *Hugh Lindsay* was of teak, 140 feet long by 25 feet wide, and powered by two 80-horsepower Maudslay engines. Mindful that the *Enterprize* had failed because of insufficient fuel supplies, Charles Malcolm saw to it that the *Hugh Lindsay* was well provisioned. She sailed from Bombay on March 20, 1830, with her hold, her cabins, and her deck so full of coal that she could hardly move. Nonetheless she ran out of coal at Aden, one third the way short of her destination. Almost a third of the journey to Suez—twelve days—was spent refueling. All the fuel had been brought by sailing ship around Africa at a cost of up to £13 per ton, and the round trip between Bombay and Suez was prohibitive at £1,700. But it was money well spent, for it ensured the *Hugh Lindsay*'s success. Whereas the *Enterprize* took as long as a sailing ship to reach India, the mail the *Hugh Lindsay* carried from Bombay reached London in a record fifty-nine days.[8]

The opening of the Red Sea Route suddenly thrust British steam power, hence British political power, into a new area of the world. To provide the *Hugh Lindsay* with a coaling station, the ruler of the island of Socotra, off the Horn of Africa, was invited to sell his island to the Presidency of Bombay in 1835; when he refused, troops seized it by force. Soon Aden was found to offer a better harbor and a healthier climate than Socotra. Its ruler, the sultan, was intimidated and bribed, but he yielded only to force in 1839.[9] The conquest of Aden, like that of Burma or the Punjab, is a classic case of secondary imperialism—a territory taken not by Britain but by British India and later, reluctantly, accepted by the British government.

Meanwhile, the Red Sea Route was flourishing. The *Hugh Lindsay* steamed fast between Bombay and Suez; one voyage took only twenty-one days. From Suez the mail went on camels' backs to Cairo, then on barges down the Nile to Alexandria. At Alexandria it waited, sometimes as long as a month, for a chance merchant ship to take it to Malta, where the British Admiralty's regular steam service began. The benefits were slim—on some early voyages two passengers and a dozen letters made the trip—and the costs were so high that the Court of Directors repeatedly forbade any further steam trips and refused to send a steamer to Alexandria to pick up the mails.

But the same public pressure that forced Parliament to appoint a Select Committee in 1834 was still strong. In 1835 the British Post Office announced it would accept letters to India via Egypt. The French inaugurated a speedy steam service from Marseilles to Alexandria, taking some traffic away from British ships. The next year Peacock, convinced of the failure of the Euphrates Route, persuaded the Directors to order two new ships, the *Atalanta,* displacing 617 tons and powered by a 210-horsepower engine, and the *Berenice,* displacing 765 tons with an engine of 220 horsepower. Seagoing steamers were now functional machines; the *Atalanta* reached India via the Cape in sixty-eight days on steam power alone.

When the members of the Euphrates Expedition returned to England in June 1837, the former governor-general of India, Lord William Bentinck, moved in the House of Commons that another Select Committee be formed "to inquire into the best means of establishing a Communication by Steam with India, by way of the Red Sea." Bentinck had been a steam enthusiast for many years. As governor-general of India from 1828 to 1834, he had introduced steamer service on the Ganges. Later, as member of Parliament from Glasgow, he had lent his name to almost every scheme for steam communication to India. To him, it was

> . . . the great engine of working [India's] moral improvement . . . in proportion as the communication between the two countries shall be facilitated and shortened, so will civilized Europe be approximated, as it were, to these benighted regions; as in no other way can improvement in any large stream be expected to flow in.

Bentinck was elected chairman of the new Select Committee, and Hobhouse became a member. The Committee discussed the merits of the steamers *Hugh Lindsay, Atalanta,* and *Berenice,* and the costs and usefulness of additional steamers on the Red Sea Route. Peacock once again was called to testify on the costs of steamer service and on the relative distances of various routes to India. The report of the Select Committee, issued July 15, 1837, only confirmed what was already obvious: the victory of the Red Sea Route.[10]

In 1838 the Court of Directors went a step further and ordered the proud old Indian Navy transformed into a steam service for mail and passengers. New steamers were soon added: the *Semiramis* in 1838, the *Zenobia* and *Victoria* in 1839, and the *Auckland, Cleopatra,* and *Sesostris* in 1840. Regular monthly steam service via Egypt was now a reality. The technological solution of the Bombay government had triumphed over the more political considerations of Palmerston and the Calcutta steam lobby.[11]

Once the Indian Navy had opened the Red Sea Route, it was not long before private enterprise joined in. The first firm to do so was the Peninsular and Oriental Steam Navigation Co., or P and O. It originated as the Dublin and London Steam Packet Co. but in 1835, after adding a service to Vigo in Spain, it became the Peninsular Steam Navigation Co. Two years later it won a contract to carry the mails to Gibraltar, the first awarded by the British government to a private firm; henceforth it was to grow in the security of lucrative and reliable mail contracts. In 1840, under the name Peninsular and Oriental, it began serving Malta and Alexandria, connecting with the Indian Navy's mail service from Bombay to Suez. In 1842 it entered the Red Sea and the Indian Ocean with a contract from Suez to Calcutta, thereby fulfilling the long-standing dream of the Anglo-Indians of eastern India. For this service it built the *Hindostan,* a ship so luxurious and so large (1,800 tons, 520 horsepower) that it made the Indian Navy's armed steamers seem cramped in comparison. The following year it offered to take over the Bombay-Suez route from the East India Company, saving the latter some £30,000 a year, but out of pride the offer was refused.

Meanwhile, passenger traffic by the Overland Route was increasing rapidly, from 275 passengers in 1839 to 2,100 in 1845; in 1847, over 3,000 travelers made the trip. And in the mid-forties, 100,000 letters were in each Overland mail. In 1845 the

P and O extended its regular service to Penang, Singapore, and Hong Kong. By the mid-fifties the once arduous and risky trip to the Orient had become fast and comparatively easy. For the Suez-Bombay route (which the East India Company finally gave up in 1854), the P and O commissioned the world's largest steamer, the 3,500-ton *Himalaya*. Speeds increased from eight or nine knots in the 1840s to eleven to fourteen knots in the mid-1850s; the whole trip from London to Bombay now took only a month. By the early 1860s the P and O's thirty-nine ships served not only India but Malaya, Singapore, China, and Australia as well. And to supply this fleet with fuel, the company employed dozens of sailing colliers and also, between Cairo and Suez, a herd of several hundred camels.[12]

## NOTES

1. G. S. Graham, "Imperial Finance, Trade and Communications 1895–1914," in E. A. Benians, James Butler, and C. E. Carrington, eds., *Cambridge History of the British Empire*, vol. 3: *The Empire Commonwealth 1870–1919* (Cambridge, 1959), p. 466.

2. Bernard S. Finn gives the following average number of days for mail to reach England in 1852: from New York twelve, Alexandria thirteen, Capetown or Bombay (via Suez) thirty-three, Calcutta (via Suez) forty-four, San Francisco or Singapore forty-five, Shanghai fifty-seven, and Sydney seventy-three. See his *Submarine Telegraphy: The Grand Victorian Technology* (Margate, 1973), p. 10. See also C. R. Fay, "The Movement Toward Free Trade, 1820–1853," in J. Holland Rose, A. P. Newton, and E. A. Benians, eds., *Cambridge History of the British Empire*, vol. 2: *The Growth of the New Empire 1783–1870* (Cambridge, 1940), pp. 412–13.

3. The race between the *Sirius* and the *Great Western* was also a race between the major figures in steamship development at that time: for the *Sirius*, Macgregor Laird, secretary of the Board of Directors of the British and American Steam Navigation Co.; and for the *Great Western*, Isambard K. Brunel, chief engineer of the Great Western Railway and later architect of the famous ships *Great*

*Britain* and *Great Eastern*. See L. T. C. Rolt, *Victorian Engineering* (Harmondsworth, 1974), pp. 85–88.

4. Harold James Dyos and Derek Howard Aldcroft, *British Transport* (Leicester, 1969), pp. 238–39; Carl E. McDowell and Helen M. Gibbs, *Ocean Transportation* (New York, 1954), p. 28; Charles Ernest Fayle, *A Short History of the World's Shipping Industry* (London, 1933), p. 241; Ambroise Victor Charles Colin, *La navigation commerciale au XIXe siècle* (Paris, 1901), pp. 39–41 and 48; Duncan Haws, *Ships and the Sea: A Chronological Review* (London, 1975), pp. 115–19; H. A. Gibson–Hill, "The Steamers Employed in Asian Waters, 1819–39," *Journal of the Royal Asiatic Society, Malayan Branch* 27 pt. 1 (May 1954):147ff; and Rolt, pp. 85–86.

5. On the relative merits of the three routes in the early nineteenth century, see Halford Lancaster Hoskins, *British Routes to India* (London, 1928), pp. 82–83, 88–89, and 105; Ghulam Idris Khan, "Attempts at Swift Communication Between India and the West before 1830," *Journal of the Asiatic Society of Pakistan* 16 no. 2 (Aug. 1971):120–21; and John Marlowe, *World Ditch: The Making of the Suez Canal* (New York, 1964), p. 41.

6. Concerning Johnston, the Calcutta Steam Committee, and the *Enterprize*, see Hoskins, pp. 86–96; Khan, pp. 119–20, 142–44, and 151–56; Gibson–Hill, pp. 122 and 134–39; Henry T. Bernstein, *Steamboats on the Ganges: An Exploration in the History of India's Modernization through Science and Technology* (Bombay, 1960), p. 32; Marischal Murray, *Ships and South Africa: A Maritime Chronicle of the Cape, with Particular Reference to Mail and Passenger Liners from the Early Days of Steam down to the Present* (London, 1933), pp. 2–3; and Auguste Toussaint, *History of the Indian Ocean*, trans. June Guicharnaud (Chicago, 1966), pp. 205–06. On the amount of the prize and the dimensions of the *Enterprize*, which are much in dispute, I have chosen to follow Gibson–Hill's version.

7. See Khan, pp. 144 and 149; Hoskins, pp. 97–98; and Daniel Thorner, *Investment in Empire: British Railway and Steam Shipping Enterprise in India, 1825–1849* (Philadelphia, 1950), pp. 23–25.

8. On the Hugh Lindsay, see Hoskins, pp. 101–09 and 183–85; Khan, pp. 150–57; Gibson–Hill, pp. 147–51; and Thorner, pp. 25–26.

9. Hoskins, pp. 123, 188–89, and 196–207; and Marlowe, *World Ditch*, pp. 33 and 42.

10. "Report from the Select Committee on Steam Communication with India; together with the Minutes of Evidence, Appendix and

Index," *Parliamentary Papers,* 1837, VI:361–617; and John Rosselli, *Lord William Bentinck: the Making of a Liberal Imperialist* (Berkeley, Calif., 1974), pp. 285–92.

11. On the triumph of the Red Sea Route, see Hoskins, pp. 109–26, 193–94, and 209–25; Gibson–Hill, p. 135; Murray, p. 8; "Biographical Introduction," in Thomas Love Peacock, *Works* (The Halliford Edition), ed. by Herbert Francis Brett-Smith and C. E. Jones, 10 vols. (London and New York, 1924–34), 1:clxx; and Carl Van Doren, *The Life of Thomas Love Peacock* (London and New York, 1911), pp. 218–19. On the *Atalanta* and the *Berenice,* see India Office Records, L/MAR/C 578: "Memorandum and Appendix on the progress of the 'Atalanta' and 'Berenice'," and 579: "Papers relating to the construction, &c. of the 'Atalanta' and 'Berenice'." On the *Semiramis,* see India Office Records, L/MAR/C 580: "Copies of various papers relating to the purchase, repairs &c. of the 'Semiramis'," and letters by "Philatmos" (pseudonym for Peacock) and "Old Canton" in *The Times,* October 31 and November 3, 6, and 7, 1838.

12. A. Fraser-Macdonald, *Our Ocean Railways; or, the Rise, Progress, and Development of Ocean Steam Navigation* (London, 1893), p. 95; W. E. Minchinton, "British Ports of Call in the Nineteenth Century," *Mariner's Mirror* 62 (May 1976):149–51; Hoskins, pp. 213–15 and 233–65; Khan, p. 151; and Thorner, pp. 33–39.

# CHAPTER NINE

# The Emergence
# of Efficient Steamships

The steamships of the 1820s and 1830s were so unprofitable to operate that their use was limited to governments, and then only when there was a pressing political need, such as communications with India. Ordinary freight and passengers continued for many years to depend on sailing ships. Even the P and O survived solely because of its mail contracts. Before the seagoing steamship could become an object of unsubsidized private enterprise, economically competitive with sailing ships on long voyages in eastern seas, several improvements had to transform it into a wholly different machine. These were the iron hull, the propeller, and the high-pressure engine.

Iron offers great advantages over wood as a shipbuilding material. It is so much stronger than wood that a two-and-a-half-inch iron girder can do the work of a two-foot oak beam. Thus an iron ship can weigh one quarter less than a wooden one of the same displacement, have one sixth more cargo space, and still be as strong. The savings in the weight of the ship, and the additional cargo space, meant that sixty-five percent of the weight of a fully loaded iron ship could be cargo, compared with only fifty percent for a wooden ship. Its lightness also enabled an iron ship to go further on less energy, a major consideration for fuel-hungry steamers.

Because of structural limitations, a ship of the best oak could hope to measure at most only about 300 feet in length. In the eighteenth century, merchantmen 150 feet long were considered large; the *Hope,* one of the largest East Indiamen, was 200 feet long by 40 in beam. Even warships were small; Nelson's *Victory* measured 186 feet by 52, hardly bigger than the *Grâce Dieu* of 1418.

In contrast, an iron ship could be almost any size. Isambard Kingdom Brunel, Britain's boldest engineer, calculated that the resistance of a hull to motion through water increases as the square of its dimensions, while its capacity increases as their cube. This led him to build ever bigger ships until in 1858 he launched the *Great Eastern.* With a length of 692 feet, she dwarfed every other ship built before the twentieth century. And if there had been engines, harbors, and customers to match her capacity, she would have proved Brunel right, for her hull fulfilled every promise.

As for shape, iron also offered advantages. A wooden ship has to be fairly stout to resist the stresses of the high seas, and the average wooden sailing ship of 1847 was 4.3 times longer than she was wide. Iron made it possible to build ships with long slim hulls, which encountered less resistance and carried more sails, and were therefore faster. Indeed the fastest sailing ships of the 1870s were built of iron, with a length six, seven, or even eight times their width.[1] Other shapes are also easier to achieve in iron than in wood, for instance the characteristic gunboat with a wide flat bottom and a moveable keel, shallow enough for rivers yet strong enough to resist ocean waves.

Iron ships are not only more cost-effective than wooden ones, but are safer as well. First, iron is stronger and more flexible than wood, therefore less likely to tear open in a collision or grounding. Second, iron ships can be built with watertight bulkheads, confining any flooding to a single compartment should the hull break open. Finally, an iron ship is less likely to burn, a consideration of some importance, especially in steamers.

Finally, iron ships are more durable than wooden ones. They are not subject to many of the ills that plague wooden ships, especially steamers: dry-rot near the boilers, bore-worms and water beetles in warm waters, and the loosening caused by the engine's vibrations. Under the best conditions a wooden ship built of good oak properly treated and dried for ten years could last as long as a century. But too often wooden ships began to rot after only two or three years, and on the average they lasted only twelve to fifteen. Iron ships, on the other hand, if regularly painted, would last until they were shipwrecked or sold for scrap.[2]

The advantages of iron ships were not apparent for a long time. The introduction of iron into shipbuilding needed a major economic stimulus to overcome inexperience and the prejudices of an age-old industry. The first iron boats were small experimental craft made in the late eighteenth century: a twelve-foot pleasure boat on the Foss River in Yorkshire in 1777, and the iron-master John Wilkinson's seventy-foot canal boat *Trial,* built in 1787. The first iron steamer, the *Aaron Manby,* came from the Horseley Iron Works in Staffordshire in 1820. Manufactured in pieces, she was transported to London and assembled on the Thames, thereby inaugurating a long era of British steamboat kits. In 1821 she steamed across the Channel and up the Seine, where she provided passenger service for the next thirty years.

Iron shipbuilding began in earnest in the 1830s, a time of unprecedented innovation in maritime matters. In 1837 the Lairds built the *Rainbow,* at 198 feet in length the largest iron steamer ever constructed. The following year the Atlantic was conquered by steam when the *Great Western* and the

*Sirius,* racing across the ocean in a record-breaking fifteen days, reached New York within hours of each other. That same year a new method of propulsion, the screw-propeller, was installed on two ships. One was Sir Francis Pettit Smith's *Archimedes;* the other, the *Robert F. Stockton* built by Laird, had a propeller designed by the Swede John Ericsson. In a tug-of-war between two sloops of equal size and power, the propeller-driven *Rattler* and the paddle-wheeler *Alecto,* the former won easily, proving the superiority of the propeller. This device was especially suited to ocean travel, where high waves made paddle-wheels plunge in and out of the water, damaging the paddles and straining the machinery. Within a decade, practically all new steamships were built with propellers.[3] And in 1839, Professor George Airy solved the compass problem, as we saw in the case of the *Nemesis.*[4] Technical objections could no longer impede the acceptance of iron steamships.

The culmination of these new technologies was Brunel's *Great Britain,* launched in 1843. This vessel combined all the latest ideas in shipbuilding. She was the largest ship ever built, she was made of iron, and had a propeller. Luxuriously fitted out, she was one of the first true ocean liners. Her career since 1843 demonstrates the perfection of iron shipbuilding in the 1840s. In 1846 she ran upon some rocks along the Irish coast that would have destroyed any other ship, yet was barely damaged. Soon repaired, the liner continued in service for another forty years. From 1886 to 1937 she served as a storage hulk, then was beached again on the Falkland Islands. Today, rescued and repaired, the *Great Britain* lies in the port of Bristol where she was built, a museum and a monument to the heroic age of steam and iron. Her hull is as strong as ever.[5]

The transition from wooden to iron ships was also the result of economic forces. During the eighteenth century and the Napoleonic Wars, Britain sacrificed her last great forests to build the ships upon which her safety and glory relied. The cost per ton of a wooden ship rose from five pounds sterling in 1600 to twenty pounds after 1775 and to thirty-six pounds in 1805. The wood alone accounted for sixty percent of the cost of a warship of that period.[6] British shipbuilders increasingly turned to foreign timber, mainly from Scandinavia and North America. Meanwhile, the young United States, with as vigorous a maritime tradition as Britain, had access to cheap abundant timber to supply its booming shipyards. By 1840 the United States had 2,140,000 tons of shipping, compared to Britain's 2,724,000 tons, and threatened to surpass the British merchant marine.[7]

It was steam and iron, pillars of the Industrial Revolution, that rescued Britain. Pig iron became cheaper as its production climbed. From £6.30 a ton in 1810 its price dropped to £4.50 in 1820, to £3.44 in 1830, and to £2.60, on the average, between 1840 and 1870. Output meanwhile rose as follows:

| | |
|---|---|
| 1800 | 200,000 tons |
| 1810 | 360,000 " |
| 1820 | 400,000 " |
| 1830 | 650,000 " |
| 1840 | 1,400,000 " |
| 1850 | 2,000,000 " |
| 1860 | 4,000,000 " |
| 1870 | 5,500,000 " |

These factors, as much as the virtues of iron, proved persuasive to shipbuilders and their customers, and prolonged the British shipbuilding and shipping supremacy for over half a century.[8]

Sometime around 1850, the iron steamer ceased being a novelty and became accepted as the standard ship. This turning point is much more elusive than the birth of an innovation like the *Aaron Manby*, or a particularly renowned example like the *Great Britain*. It may have come in 1848 when the Penin-

sular and Oriental, tired of the termites that plagued its
wooden ships, bought its first iron steamers for the Indian
Ocean. Or perhaps it came in the mid-fifties. Until then the
British Post Office insisted that the mails be carried on wooden
paddle-steamers, and in deference to its wishes the Cunard
Line waited until 1856 to buy its first iron ship, the *Persia*.
Also in 1856, Lloyds of London issued specifications for iron
merchantmen. When such conservative institutions as these
finally accept an innovation, we can be sure it has truly
arrived.[9]

The best steamships of the 1840s—the *Great Britain*, for example—were in almost all respects closer to the great ocean
liners of the twentieth century than to the *Clermont* or the
*Hugh Lindsay;* only their engines were still relatively rudi-
mentary. Since the beginning of the nineteenth century, engineers had known how to make engines more efficient by increasing the pressure of the steam. In fact, stationery land
engines, such as those pumping water from the tin mines of
Cornwall, were more efficient than any marine engine. The
trouble with marine engines was that they used seawater,
which left salt deposits in their boilers. In high-pressure engines these accumulations were extremely dangerous. As a result, marine engines of the 1850s were limited to a pressure of
about 25 pounds per square inch. In 1834, however, Samuel
Hall had invented a method of recycling steam back to the
boilers in the form of salt-free distilled water; this saved boilers from corrosion and allowed higher pressures. Hall's surface condenser was gradually improved upon until the late
1850s, when it was incorporated into almost all new seagoing
steamers.[10]

The high pressures allowed by surface condensers in turn
prompted a device that utilized this pressure more effectively:
the compound engine. This system, patented by Charles Rudolph and John Elder, used two cylinders—a small one fed

high-pressure steam from the boiler and a larger one that was filled with the low-pressure steam exhausted from the small one. Though this device was introduced as early as 1854, it did not come into its own until the 1860s. In 1862, Alfred Holt, a civil engineer, built the *Cleator* specifically for long-distance trade in eastern seas. Three years later he launched three more compound-engine steamers, the *Agamemnon,* the *Ajax,* and the *Achilles,* with which he founded the Ocean Steamship Co., sailing to the Indian Ocean and the Far East. In contrast to the *Britannia* of 1843 whose engine had a pressure of 9 pounds per square inch and burned 5 pounds of coal per horsepower per hour, Holt's ships had pressures of around 60 psi and a coal consumption of 2.25 lbs/hp/hr. In 1865 one sailed from England to Mauritius, 8,500 miles away, without refueling. Rudolph and Elder, meanwhile, were building compound-engine ships for their Pacific Steam Navigation Co. which served the Far East. At last seagoing steamships were economical enough to interest private shippers carrying ordinary cargoes.[11]

## NOTES

1. Maurice Daumas, ed., *Histoire générale des techniques,* 3 vols. (Paris, 1968), 3:328.

2. On the advantages of iron over wooden ships, see Bernard Brodie, *Sea Power in the Machine Age: Major Naval Inventions and their Consequences on International Politics, 1814–1940* (London, 1943), pp. 149–54; Daumas, 3:359; Cammell Laird & Co. (Shipbuilders & Engineers) Ltd., *Builders of Great Ships* (Birkenhead, 1959), p. 12; Ambroise Victor Charles Colin, *La navigation commerciale au XIXe siècle* (Paris, 1901), pp. 55 and 60–61; Harold James Dyos and Derek Howard Aldcroft, *British Transport* (Leicester, 1969), p. 239; Charles Ernest Fayle, *A Short History of the World's Shipping Industry* (London, 1933), p. 240; Duncan Haws, *Ships and the Sea: A Chronological Review* (New York, 1975), p. 117; L. T. C. Rolt, *Victorian Engineering* (Harmondsworth, 1974), pp. 85–86; David B. Tyler, *Steam Conquers the Atlantic* (New York, 1939),

p. 113; René Augustin Verneaux, *L'industrie des transports maritimes au XIXe siècle et au commencement du XXe siècle,* 2 vols. (Paris, 1903), 1:304 and 2:4–8; and Halford Lancaster Hoskins, *British Routes to India* (London, 1928), p. 81.

3. W. A. Baker, *From Paddle–Steamer to Nuclear Ship: A History of the Engine–Powered Vessel* (London, 1965), pp. 10–12; Colin, pp. 41–43; Haws, p. 130; and Verneaux, 2:44–45.

4. See Chapter 2, note 11.

5. On the earliest iron ships, see James P. Baxter, III, *The Introduction of the Ironclad Warship* (Cambridge, Mass., 1933), p. 33; Charles Dollfus, "Les origines de la construction métallique des navires," in Michel Mollat, ed., *Les origines de la navigation à vapeur* (Paris, 1970), pp. 63–67; Haws, p. 117; George Henry Preble, *A Chronological History of the Origin and Development of Steam Navigation,* 2nd ed. (Philadelphia, 1895), pp. 119–35; Hereward Philip Spratt, *The Birth of the Steamboat* (London, 1958), p. 40; Tyler, pp. 112–13; and Verneaux, 2:8–9. On the *Great Britain,* see Rolt, pp. 89–91 and Carl E. McDowell and Helen M. Gibbs, *Ocean Transportation* (New York, 1954), p. 27.

6. Daumas, 3:359–60.

7. Colin, pp. 2–4; Dyos and Aldcroft, pp. 235–36.

8. On the economics of iron, see Charles K. Hyde, *Technological Change and the British Iron Industry 1700–1870* (Princeton, N.J., 1977), pp. 137–39, 163, 170, 234, and 245. See also Brodie, pp. 131–33.

9. Haws, p. 142; Cammell Laird, p. 18; and Verneaux, 2:61–62.

10. On surface condensers, see A. Fraser-Macdonald, *Our Ocean Railways; or, The Rise, Progress, and Development of Ocean Steam Navigation* (London, 1893), p. 213; Rolt, pp. 96–97; "Ship," in *Encyclopaedia Britannica* (Chicago, 1973), 20:407; and Daumas, 3: 355–56.

11. On the compound-engine ships of the 1850s and 1860s, see: Colin, pp. 48–49; Dyos and Aldcroft, pp. 238–40; Fraser-Macdonald, pp. 213–14 and 225; Haws, pp. 120, 142, 149, and 157; Francis E. Hyde, *Liverpool and the Mersey: An Economic History of a Port 1700–1970* (Newton Abbot, 1971), pp. 53–54; Adam W. Kirkaldy, *British Shipping: Its History, Organization and Importance* (London and New York, 1914), pp. 90–101; McDowell and Gibbs, p. 28; Thomas Main (M.E.), *The Progress of Marine Engineering from the Time of Watt until the Present Day* (New York, 1893), pp. 56–66; "Ship," 20:409; Verneaux, 2:36–39; and Roland Hobhouse Thornton, *British Shipping* (London, 1939), p. 66.

# CHAPTER TEN

# The Suez Canal

Some technologies, like steam or iron, are born in obscurity, and only gradually change their environment as they evolve. Others on the contrary have the quality of a long-heralded millennium. No event of the nineteenth century was awaited with such fervent expectation, or celebrated with such drama and enthusiasm, as the opening of the Suez Canal. On November 17, 1869, Empress Eugénie of France entered the canal on board her imperial yacht *Aigle* for the three-day journey to Suez. The emperor of Austria, the crown prince of Prussia, the grand duke of Russia, and a host of other dignitaries, publicity-seekers, journalists, and party-goers followed on board sixty-eight steamers. Champagne flowed as speeches filled the air. It was the social event of the century, and it cost a staggering £1,300,000.

The story of the canal is filled with ironies. While there were canals from the Red Sea to the Nile in Pharaonic times, the first to propose a Mediterranean–Red Sea canal was the eighth-century caliph Harun al-Rashid. Napoleon Bonaparte, during his invasion of Egypt in 1797–98, ordered his engineers to survey the proposed canal route. A mistake on their part—

they concluded that the Red Sea was thirty-two inches higher than the Mediterranean—postponed the project for several decades.

The Red Sea Route pioneered by the *Hugh Lindsay* made plain the need for a canal. In an otherwise easy journey, the passage through Egypt quickly proved to be the most serious bottleneck along the route to the East. Travelers had to spend eight to ten days of misery crossing the desert, camping out, or waiting in one of Suez's infamous squalid hotels. In addition, all the coal for the steamers to India had to be brought to Suez on the backs of camels.

In the 1830s the explorer Francis Chesney and the French engineer Linant de Bellefonds recognized the error in their predecessors' calculations and proposed to the pasha of Egypt, Mehemet Ali, a sea-level canal. In 1841, in the midst of a war between Britain and Egypt, the managing director of the P and O, Arthur Anderson, journeyed to Egypt to study the Overland transit situation. He was graciously received by Mehemet Ali, to whom he mentioned the need for a canal. In 1846–47, Prosper Enfantin, a disciple of the positivist Count de Saint-Simon, founded the Société d'Etudes pour le Canal de Suez. Several teams of engineers surveyed the route of the proposed canal, and all reached the same conclusions: The Red Sea and the Mediterranean were at the same level, and between them lay a flat land of sand dunes and salt marshes offering no serious engineering difficulties to canal builders. Only political obstacles prevented the Suez Canal from being completed twenty years earlier than it was.

Mehemet Ali was lukewarm to the idea of a canal, and his nephew Abbas, who succeeded him (1849–54), disapproved of it. Not until Mohammed Saïd became pasha did the project begin. Saïd was a close friend of the French consul, Ferdinand de Lesseps, who hoped to carry out the plan proposed by Linant de Bellefonds and Prosper Enfantin. On November 30, 1854, Saïd gave de Lesseps the concession to build the canal.

A major obstacle remained, however: Palmerston's interpretation of the Near Eastern question. Palmerston recognized that a Suez Canal would be a second Bosporus, a waterway of strategic importance to all the great powers. If it were built, Egypt would break away from the Ottoman Empire and fall into the orbit of whatever European nation controlled the canal. But Britain needed to support the Ottoman Empire against the ambitions of Russia. Furthermore, a Frenchman was directing the project, and since the French government seemed to favor it, the British government therefore had to oppose it.

This policy was most unpopular with the British business world. Shipbuilders, cotton manufacturers, the P and O, the East India Company, and the *Calcutta Review* (the voice of the Anglo-Indian community) all supported the canal. The tardiness of the British government to respond to the Indian Mutiny of 1857 fueled their arguments. Year by year the government's position grew weaker. As a palliative, it offered substitutes. One was the Cairo-Suez Railroad, built in the 1850s to the great relief of passengers, but of little consequence for freight. The other was the projected Euphrates Valley Railroad, a reincarnation of the old Mesopotamian Route and, like Chesney's dream twenty years earlier, just as much a failure.

If the canal was the most logical solution, it did not emerge victorious without a long struggle. Gradually de Lesseps inched his way toward his goal. In 1856 he carried out a great public relations campaign that gained him the support of the British business and shipping circles. He traveled incessantly between Cairo, Constantinople, Paris, and London, conducting a one-man diplomatic mission. In the end, of the 300,000,000 francs (£12,000,000) spent on the canal, one third went for political, promotional, and incidental expenses.

In 1859 de Lesseps sensed that the opposition had weakened enough to tolerate a *fait accompli*. So he founded the Compagnie Universelle du Canal Maritime de Suez, mostly with French capital, and began construction. Digging a canal from the Mediterranean to the Red Sea was well within the engineering knowledge of the time; indeed all over the world harbors and ships' channels were being deepened and enlarged to accommodate the ever-increasing size and number of ships. The difficulties lay in the magnitude of the task and in the environment.

Until 1864 the digging was done by *corvée* laborers supplied by the Egyptian government under its contract with the canal company. During the most active phase of construction, 20,000 or more men worked for a period of one month, then were replaced by another 20,000. Including time spent traveling to and from the canal site, these men were absent from their homes for about three months. At the construction site there were huge logistical problems. During the first years, fresh water had to be carried in from the Nile; 3,000 camels and donkeys were employed in this task alone. Nonetheless the men often went thirsty and suffered outbreaks of dysentery and cholera. After 1862 a Sweet Water Canal from the Nile reached Ismaïlia, midway between Suez and the Mediterranean, alleviating this problem.

Tools were imported from Europe, causing, in some cases, culture shock in a wheelless society:

> At first there was some difficulty in getting the *indigènes* to use the wheel-barrow; so much so, that some commenced by carrying them on their heads. They were in the habit of using either a small basket, holding only a few handfuls of earth; or one shovelled it into a sack, whilst another carried it away.[1]

More difficult even than digging the canal was building a new harbor at the northern terminus. Situated on a shallow shifting sand beach, Port Saïd required two jetties, one of them two miles long, out into the sea. To build the jetties and the city itself, stone was brought by ship from the nearest quarry, 150 miles to the west beyond Alexandria. Later 327,000 cubic yards of concrete blocks, made locally from canal dredgings, were dumped into the sea to protect the harbor from storms and currents.

As construction proceeded, the British government, which had always viewed the canal with suspicion, began vehemently to denounce the use of *corvée* labor. Behind its humanitarian indignation at the involuntary servitude of the peasantry, critics noted, lay a concern also for the Egyptian cotton crop which Lancashire industry sorely needed during the "cotton famine" of the American Civil War. Whatever its motives, British pressure on the Ottoman and Egyptian governments forced a technological revolution upon the canal company.

By 1863 workers had dug a narrow channel, just big enough for mechanical dredgers, most of the way across the isthmus. At that point, to replace the *corvée* labor that was phased out in June 1864, the company ordered machines, which began arriving a year later. There were dredges with iron buckets pulled by an endless chain around two drums, one at the end of a movable arm, the other attached to a framework over the hull; the largest of these dredges were 110 feet long, had seventy-five horsepower engines, and cost £20,000 apiece. To carry away the dredgings the French engineer Lavallay, one of de Lesseps's contractors, designed two new pieces of machinery. The *long couloir* was a duct 246 feet long sloping downward away from the dredge, along which dredgings were swept by a flow of water and a series of paddles pulled by a chain. For locations where the canal banks were too high for the *long couloir,* he designed an *élévateur,* a 164-foot-long inclined plane along which boxes of dredgings were pulled by a chain, and their contents dumped off the high end. Tramways were built for dry excavating, their cars pulled by stationary steam engines or donkeys.

By late 1868 these machines had cost the phenomenal sum of £2,400,000. Every month they burned £20,000 worth of coal and produced over 10,000 horsepower, to excavate up to

2,600,000 cubic yards of earth. Altogether, 78,500,000 cubic yards were dug by machine in 1865–69, compared to 20,000,000 by hand from 1861 to 1864. Construction by massive manpower, an Egyptian tradition since the first pharaohs, had given way to the biggest concentration of mechanical energy ever assembled.

As the work neared completion, the Ottoman government finally withdrew its opposition, as did the British. By 1869 British diplomacy was actively seeking to neutralize the canal and guarantee its safety.

The Suez Canal drastically cut the distance between Europe and the East. Compared to the Cape Route, the voyage from London to Bombay was shortened by fifty-one percent, to Calcutta by thirty-two percent, and to Singapore by twenty-nine percent. Its most important impacts were upon the East-West trade and shipbuilding. The first decade of the canal's operation was difficult, for sailing ships could not use it, and there were as yet few steamers equipped for the long voyage east. By 1882, however, the canal was operating at capacity. The introduction of electric headlights on ships after 1887 allowed night travel, cutting the transit time in half. Several times the canal had to be straightened, widened, and deepened to accommodate the increasingly larger and more numerous ships that passed through it. The following table shows the growth of this traffic:[2]

| Year | Number of Ships | Tons Capacity |
|------|-----------------|---------------|
| 1870 | 486 | 436,609 |
| 1875 | 1,494 | 2,009,984 |
| 1880 | 2,026 | 3,057,422 |
| 1885 | 3,624 | 6,335,753 |
| 1890 | 3,389 | 6,890,094 |
| 1895 | 3,434 | 8,448,383 |
| 1900 | 3,441 | 9,738,152 |

The Suez Canal, hinge of East and West, was a global achievement. Built mostly with Egyptian labor and French money and machinery, it served mainly the interests of Great Britain. The first ship to pay a toll on the canal was British, and within a few years, three quarters or more of the ships that used it were of British registry. As the canal began to make a profit, the Khedive Ismaïl sank deeper into debt. In 1875 he sold his 176,602 shares in the canal company to the British government for £3,976,582. Four years later, bankrupted by his wasteful tastes and the generosity of the Western powers in lending him money, he was deposed by the Ottomans. In 1882, Britain added Egypt to her empire in the name of financial solvency. The logic of the Red Sea Route to India had overcome at last those long-lived British compunctions about territorial aggrandizement.[3]

## NOTES

1. J. Clerk, "Suez Canal," *Fortnightly Review* 5 (New Series, 1869): 207. For a reasoned explanation of the nonuse of the wheel in Arab lands see Richard W. Bulliet, *The Camel and the Wheel* (Cambridge, Mass., 1975).

2. René Augustin Verneaux, *L'industrie des transports maritimes au XIXe siècle et au commencement du XXe siècle*, 2 vols. (Paris, 1903), 1:329. Halford Lancaster Hoskins, *British Routes to India* (London, 1928), pp. 447, 469, and 473, gives slightly different figures.

3. On the construction of the Suez Canal, see Clerk, pp. 80–100 and 207–25; John Marlowe, *World Ditch: The Making of the Suez Canal* (New York, 1964); Lord Kinson, *Between Two Seas: The Creation of the Suez Canal* (London, 1968), especially pp. 149–225; and Hoskins, pp. 292–315, 345–72, 400–407, and 447–79.

CHAPTER ELEVEN

# The Submarine Cable

It has become a commonplace to characterize our own time by its "information explosion." The goal of higher education is no longer erudition (how to store small amounts of information) but research methodology (how to acquire and handle large amounts). The speedy acquisition of data has become as central to the concerns of business as labor, machines, and raw materials. As governments sink into ever deeper bondage to their intelligence services, the media-struck public awaits the news each day, like investors at their ticker-tape machines.

Addiction to data and speed are nothing new, of course. The need for rapid information was one of the forces behind the opening of steam communication with India. But it went even further, creating a network of telegraph wires and submarine cables that transmitted messages over global distances at hitherto unimaginable speeds. These cables were, in the words of the historian Bernard Finn, "the grand Victorian technology."[1]

Land telegraphy had been a reality for two decades before underwater telegraphy came of age. This lag was not the result of a lack of desire or money, but of technology. Until the 1860s the ocean depths were unknown, and the physics of transmitting electrical impulses over thousands of miles was in its infancy. It took both scientific research and costly trial and error to solve these problems.

The first attempt at underwater telegraphy seems to have been a cable laid across the Hooghly River in Calcutta in 1839. Not until 1843, however, was a good insulating material discovered; this was gutta-percha, a natural plastic formed from the sap of a Malayan tree. In 1850, John and Jacob Brett laid a cable across the English Channel. It was made of a single strand of copper wire coated with gutta-percha. A few hours after its debut, a fisherman's anchor broke it. The next year another cable was laid alongside the first. This one was made of four copper wires insulated with gutta-percha, sheathed with iron wire, and then wrapped in jute and coated again with pitch; thus protected, it worked well into the next century. Two years later, in 1852, a cable linked England and Ireland. Submarine telegraphy was born.

The investing public, easily carried away by gold rushes and railway booms, seized upon the new technology with undue haste. If a cable could span the Channel, then surely a longer piece of the same cable could span the Atlantic Ocean, thought the speculators. In 1857 and 1858 two steamers stretched a cable from Britain to America. A few messages were sent, including one cancelling an order for two regiments of troops to be sent from Canada to India, saving the British government £50,000. Then the line died out.

A similar fate befell the first submarine cable to India. The first scheme to link Britain with India was formulated by the European and Indian Junction Telegraph Co. Ltd., founded in 1856 with the intention of laying a land line to the Persian Gulf and a submarine cable thence to Karachi. At the time of the Indian Revolt of 1857, the 4,500 miles of land lines in India helped the British move troops quickly and crush the uprising in a few months. The question of a telegraphic link with Britain therefore became a major issue, receiving the attention of the House of Commons and of Palmerston himself. In 1856 the Red Sea and India Telegraph Co. was founded to link Constantinople with Alexandria, and Suez with Aden and Karachi. In 1859 the first Red Sea cable was laid. It was a single strand of copper wire coated with gutta-percha and covered with hemp, and it weighed one ton per mile. It was laid by crude machines that pulled it taut so that it hung between underwater peaks. In places terero bore-worms ate through the insulation, and in other places the cable got so heavily encrusted with growths that it broke under their weight. The first cable to India, which cost £800,000, never transmitted a single message.

The disasters of the Atlantic Ocean and Red Sea cables stirred the British government into action. The Board of Trade and the Atlantic Telegraph Co. appointed a joint committee to study submarine telegraphic technology. William Thomson (later Lord Kelvin) advised the Atlantic cable project; he designed instruments for sending and receiving messages and invented methods of deep-sea sounding. The questions surrounding cable construction and laying methods were studied. When the joint committee finished its work in September 1860, the basic techniques of cable manufacture and laying and of electrical transmission had been perfected. The telegraph, at last, could reach across the seas and oceans.

There followed a rush to install submarine cables. In 1861 the French laid one from Port-Vendre in southern France to Algiers, and the British laid one from Malta to Alexandria, both with a clear colonial role. From 1862 to 1864 a new submarine cable joined Karachi with the Persian Gulf; this one weighed four times more than its precursor and worked well. At the gulf, it connected with a land line through Persian and Ottoman territory to Constantinople; from there several land lines radiated to Europe. In 1865, a year before the first cable to America, the telegraph linked Britain with India.

If the very existence of this line demonstrated the importance of India in British eyes, its performance fell far short of its imperial purpose. Between London and Constantinople, the lines crossed various states whose own official telegrams took precedence over foreign messages. It also grated on British sensibilities that their telegrams could be spied on, altered, or postponed at will by Continental officials. Worse was the fate of telegrams beyond Constantinople. There, long lines across poorly controlled territories often broke down. Copper wires were stolen. Turkish and Persian clerks, with a less than perfect command of the English language, garbled the messages at the many relay stations along the way where telegrams had to be taken down and retransmitted. As a result telegrams between Britain and India often took a week, and occasionally a month or more, to reach their destination.

As in all crises relating to the empire, the House of Commons appointed another Select Committee in 1866 "to inquire into the practical working of the present system of Telegraphic and Postal Communications between this Country and the East Indies." Two new lines to India were laid. One, the Indo-European Telegraph Co., was a British-owned land line across Belgium, Germany, Russia, and Persia to the British submarine cable in the Persian Gulf. The other was a completely British submarine cable to Gibraltar, Malta, and Alexandria, and from Suez to Aden and Bombay; only from Alexandria to Suez was there a land line, with British operators. In 1870, at last, Britain and India were linked by an efficient telegraph line that regularly conveyed messages in about five hours.

In the 1870s, after the triumph of the cables to America and India, there emerged a powerful submarine cable industry. Aided by further technological improvements and the active policy of the British government, it laid cables around the world. Curb transmission, patented in 1861, sharpened the signal by sending a reverse pulse immediately after the main pulse. Duplex telegraphy, introduced in the mid-seventies, al-

lowed messages to be sent in opposite directions simultaneously over the same line, thus doubling its capacity. The si-
phon recorder and other automatic machines replaced the
human hand with punched tape. These and other improvements sped up telegrams and reduced their cost. A twenty-word
message on the first telegraph line to India cost 101 shillings
and took several days; by the end of the century such a
message cost four shillings and took half an hour. And the
customers responded; whereas a few dozen telegrams were
sent in 1870, two million were transmitted in 1895. The new
medium not only existed, it penetrated and transformed the
daily routines of Anglo-Indian relations.

Soon dozens of other cables were laid throughout the seas
and oceans. In 1871 a cable linked India with Penang, Saigon,
Hong Kong, and Shanghai, and another went from Singapore
to Java and Port Darwin; Europe was now in contact with
Southeast Asia, China, and Australia. Two years later cables
linked Europe with Brazil and Argentina, and in 1875 a cable
reached the west coast of South America. In 1876 an extension
of the Australian cable reached New Zealand. In 1879 another
was laid along the east coast of Africa from Aden to Zanzibar
and Mozambique, reaching South Africa. Finally in 1885 two
cables along the west coast of Africa linked Europe with West
and South Africa. All the while, along the earliest cable routes
in the North Sea, the Atlantic Ocean, and the Mediterranean,
new cables were added alongside the old as demand rose; for
example, in 1900 there were seven cables between France and
Algeria.

The earliest cables were laid by a number of entrepreneurs
or hastily formed companies. The seventies saw the consolidation of these groups into one giant monopoly. In 1864, John
Pender, a Manchester merchant, brought together the Gutta
Percha Co. and Glass Elliott and Co. into the Telegraph Construction and Maintenance Co., or T C and M. In 1866 he
took shares in the Atlantic Telegraph Co., which laid the ca-

ble to America that T C and M had manufactured. By 1870, Pender was chairman of the British Indian Submarine Telegraph Co. Two years later he founded the Eastern Telegraph Co., a firm that soon became the Eastern and Associated Companies, or "Electra House Group," controlled by the Pender family. Of the approximately 190,000 miles of submarine cables in the world in 1900, seventy-two percent were British-owned, mostly by Eastern and Associated, as several foreign authors ruefully noted.[2]

Though owned by private investors, the cable network of the seventies was no triumph of free enterprise, but the beneficiary of governmental munificence. The first windfall came in 1870 when the British government nationalized its domestic telegraph companies, releasing capital for investment in submarine cables. Later it subsidized cable companies either annually or in a lump sum; for instance, it paid £1,100,000 to the Eastern and South African Telegraph Co. to lay the Aden–South Africa cable, and £19,000 a year to the Africa Direct Telegraph Co. for its West African line. Just as mail contracts kept the Peninsular and Oriental and the Cunard lines prosperous in good times and bad, so did subsidies support the cable firms. Not once between 1873 and 1901 did Eastern make less than 6.75 percent profit per annum.

The cable network of the seventies was composed of economic cables, that is to say, lines that were useful to business and private customers. After 1880 such opportunities were depleted, and a new era opened. The Admiralty and the Colonial, War, and Foreign offices had gotten accustomed to communicating by telegram and wished to extend that possibility to all parts of the British Empire. As jingoist sentiments rose, it became increasingly galling to have British telegrams cross non-British territories. Thus, ever more distant lines with ever less economic value were laid for political reasons. One such cable was the Suez–Suakin line laid as part of the British invasion of Egypt in 1882. Another was a direct cable from Britain to the Cape of Good Hope in 1899–1901, which was used in the Boer War. Under Admiralty pressure this line was extended to Mauritius and from there to Ceylon, Singapore, and Australia in 1901–02. While these lines duplicated the Eastern cables, they did not pass through Egypt or the Red Sea, and therefore were thought to be strategically safer.

The epitome of strategic cables was the "All-Red Route," a scheme to gird the globe with a cable passing only through British territories. By the 1890s there were several cables to Canada and Australia. The missing link was a cable from British Columbia to New Zealand. Since not even so patriotic a firm as Eastern and Associated would get involved in any scheme so patently unprofitable, the plan was handed to the Pacific Cable Board, a consortium owned by the governments of Great Britain, Canada, New Zealand, and Australia. In 1902 this line was completed, and all parts of the British Empire could henceforth communicate by a cable network upon which the sun never set.[3]

Cables were an essential part of the new imperialism. At the rudest level, they gave value to a handful of mostly deserted islands in the most isolated parts of the world: Ascension, St. Helena, Norfolk, Rodriguez, Fanning, and Cocos. Like Guam and Midway for the United States, these islands served as relay stations for Britain. In a few instances, cables helped the empire to expand: South Africa in 1879 and 1901, Egypt in 1882, and West Africa in 1885 all came under Britain's wing. But more important, cables served to tie the European empires together. In times of peace they were the lifelines of the ever-increasing business communications that bound imperialist nations to their colonies around the world. In times of crisis, they were valuable tools of diplomacy; in the Fashoda confrontation of 1898, Kitchener communicated with London by way of an underwater telegraph cable sunk in the Nile, while his French opponent Marchand was cut off from Paris.[4] And in times of war, the cables were security itself. In World War One the British and French empires, held together by the telegraph, supplied their metropoles with troops, food, and raw materials. And the German Reich learned in August 1914, when its cables were cut, that the world communicated on British sufferance.

## NOTES

1. Bernard S. Finn, *Submarine Telegraphy: The Grand Victorian Technology* (Margate, 1973). See also Frank James Brown, *The Cable and Wireless Communications of the World; a Survey of Present Day Means of International Communication by Cable and Wireless, Containing Chapters on Cable and Wireless Finance* (London and New York, 1927); E. A. Benians, "Finance, Trade and Communications, 1870–1895," and Gerald S. Graham, "Imperial Finance, Trade and Communications, 1895–1914," in E. A. Benians, James Butler and C. E. Carrington, eds., *Cambridge History of the British Empire,* vol. 3: *The Empire-Commonwealth 1870–1919* (Cambridge, 1959), chs. 6 and 12, respectively; August Röper, *Die Unterseekabel* (Leipzig, 1910); and Willoughby Smith, *The Rise and Extension of Submarine Telegraphy* (London, 1891). On cables to India, see Halford Lancaster Hoskins, *British Routes to India* (London, 1928), pp. 373–97.

2. See Ambroise Victor Charles Colin, *La navigation commerciale au XIXe siècle* (Paris, 1901), p. 147; and Röper, p. 85.

3. On the strategic value of cables, see P. M. Kennedy, "Imperial Cable Communications and Strategy, 1870–1914," *English Historical Review* 86(1971):728–52.

4. Richard Hill, *Egypt in the Sudan 1820–1881* (London, 1959); and Kennedy, p. 728.

CHAPTER TWELVE

# The Global Thalassocracies

While the Suez Canal and the cables marked dramatic turning points, the world communications network also benefited from the more gradual evolution of shipbuilding and shipping. On the engineering side, the main improvements were the introduction of steel hulls and triple-expansion engines. The economics of shipping, meanwhile, led to ever larger and more specialized vessels, to improvements in harbors and other navigational infrastructures, and to organizations designed to make the most efficient use of these means.

The introduction of steel in shipbuilding followed naturally in the wake of the switch from wood to iron and involved none of the moral and social qualms of the earlier transition. Because steel was stronger than wrought iron, a steel ship could be built fifteen percent lighter than an iron ship of equal strength and dimensions. This in turn brought about higher speeds and lower fuel consumption.

The first steel-hulled vessel was the little river steamer *Ma Roberts,* which Laird built for Livingstone's Zambezi River Expedition in 1858. She gave poor service, unfortunately, for her hull rusted badly and her engine was too weak. She finally was wrecked on the banks of the Zambezi.[1] Next came the *Banshee,* a blockade-runner in the American Civil War, and a number of other steel warships designed for speed, not economy. Only in the late 1870s, after the price of steel had become competitive with that of wrought iron, did steel replace iron in the construction of merchantmen.[2] The first large steel ocean liner was the 1,777-ton *Rotomahana,* built for the Union Steamship Co. of New Zealand in 1879. After the P and O found that its steel-hulled *Ravenna* gave better service than iron-hulled ships, it ceased ordering the latter. By 1885, forty-eight percent of all new steamers were built of steel, and by 1900, all but five percent were similarly constructed.[3]

Just as steel was the natural successor to iron, so was the triple-expansion engine the successor to the compound engine. The use of steel in the manufacture of boilers, pipes, and engine parts allowed engineers to increase the steam pressure from 60 psi in the 1860s to 150 psi in the late seventies, then to 200 psi in the nineties. At these higher pressures, steam did not lose all its expansive power even after passing through a second cylinder; it was to capture this remaining energy that engineers added a third cylinder.

The resulting engine ran faster and more smoothly on considerably less fuel than a compound engine of the same power. Liners of the 1860s and 1870s had burned 1.75 to 2.50 pounds of coal per horsepower per hour. The *Aberdeen,* the first large triple-expansion-engine ship, burned only 1.25. This ship, especially well suited to very long routes, set a record by sailing from Plymouth to Melbourne in forty-two days. Later triple-expansion engines burned as little as one pound of coal per horsepower per hour. Put another way, the energy contained in one sheet of paper, if burned in such an engine, could move a ton of freight one mile. The triple-expansion engine was to be (along with a few quadruple-expansion engines) the last and most perfect expression of the Newcomen-Watt reciprocating steam engine that had so revolutionized the world of the nineteenth century. By World War One, however, two newcomers, the steam-turbine and the diesel engine, had begun to replace it. Such was the pace of maritime progress.[4]

Shippers, ever on the lookout for greater efficiencies, had not forgotten Brunel's revelation that a big ship can carry cargo at a lower cost than several small ones (other things being equal). In the 1850s Brunel had badly miscalculated the world's demand for his *Great Eastern;* the world of the nineties, however, was quite ready for such large ships. The largest were the Atlantic liners, stars of the ocean, whose story has often been told. But in the humbler trades also, every decade saw larger ships. In the 1850s a steamer of 2,000 tons was considered large. By the 1890s ships of 6,000 to 8,000 tons were quite common. The average P and O liner in 1860 displaced 1,490 tons, and the largest 2283 tons; in 1897 the average increased to 4,896 tons, and the largest displaced 8,000. By 1914 steamers of 20,000 tons or more were not unusual.[5]

Steamers grew in number as well as size. The maritime historian Adam Kirkaldy gives the following figures on the tonnage of the world's steamships:[6]

| | |
|------|-------------------|
| 1850 | 700,000 tons |
| 1860 | 1,500,000 " |
| 1870 | 2,700,000 " |
| 1880 | 5,500,000 " |
| 1890 | 10,200,000 " |
| 1900 | 16,200,000 " |
| 1910 | 26,200,000 " |

The increased efficiency of steamers, the competition resulting from the greater number of ships and the shortened distances through the Suez Canal all contributed to a significant decline in freight rates during this period. According to the maritime historian A. Fraser-Macdonald, the cost of shipping a ton of cargo from Bombay to England fell from ten or twelve pounds sterling in 1869 to twenty or thirty shillings in 1893, a ninety-percent decline; rates on longer routes fell even more.[7] Other authors cite less extreme declines but all agree that the rates dropped at least sixty percent, and that much of this decline occurred because of lower insurance and handling charges on goods carried in safe and reliable steamships.[8]

The growth in the size and number of ships was paralleled by a great expansion in trade. Between 1860 and 1910, Britain's trade with India increased threefold, from £38,600,000 to £116,100,000; with Australia from £17,100,000 to £60,000,000; and with South Africa from £3,900,000 to £29,900,000, and all during an era of gradually falling prices.[9] The Suez Canal, forced to accommodate ships that were, on the average, over three times larger than those for which it had been designed, was deepened and widened, so that large ships could pass through simultaneously. By 1914 all important seaports had a minimum depth at dockside of thirty-six feet. Harbors deep and spacious enough to handle the new ships were enlarged and protected with concrete breakwaters. Among the most important ports of the post-Suez era were brand-new colonial cities such as Karachi, Mombasa, Singapore, Port Saïd, and Aden. Older cities like Shanghai and Bombay were transformed beyond recognition. By 1892, Hong Kong, the little island wrested from China in the Opium War, cleared more shipping than Liverpool and almost as much as London.[10]

All the world's seaports sold coal to passing steamers, but some were almost exclusively coaling stations and ports-of-call, with little other trade: Port Saïd, Mauritius, Acheen, Las Palmas, St. Vincent, and Papeete are a few examples. Others, like Aden, Djibouti, Singapore, and Honolulu, also became important naval bases.[11]

When steamers conquered the seas, it was not at the expense of sailing ships but with their assistance. The early years of the ocean liners coincided with the greatest age of sailing ships, brought close to perfection by their rivalry with steam. In their pursuit of speed through harmony with the forces of nature, naval architects of the mid-century created the China clippers, triumphs of aesthetics and technology, the cathedrals of the sea.[12] Such was the increase in maritime trade in the forties and fifties that each type of vessel had a role to play, and a growing one. While steamers took over the North Atlantic passenger service and the mail service to India, sailing ships carried the mundane freight traffic of other seas.

The Suez Canal ended the clipper era, but it did not banish sails from the oceans of the world. Sailing ships held their own in two types of trade until the end of the century. One was carrying bulky cargo from areas most remote from the world's coal mines: Australian wool and wheat, Indian rice and jute, grains from the American West Coast, and nitrates from Chile. The other—and more significant—was the outbound shipment of coal to satisfy the fuel needs of steamers on all the world's oceans. Even after coal was discovered in India, South Africa, Japan, and elsewhere, ships still paid a premium for the higher-quality Welsh anthracite. In 1879, Britain exported 11,703,000 tons of coal, and in 1895, 33,101,000 tons, much of it on sailing ships.

The turn of the century, however, marked the end of the sailing era. In 1875, 56,537 out of 61,902 ships in Lloyd's Register (ninety-one percent) were sailing ships; their tonnage represented seventy-two percent of the figure for the world. By then Britain was already building more steam than sailing tonnage. In 1899–1900, sixty percent of all ships (22,856 out of 38,180) were propelled by the wind, but their tonnage represented only one quarter of the world's total (6,795,782 tons out of 27,673,528).[13]

Two changes doomed the sailing ship after the 1880s. One was the greater fuel efficiency of steamers brought about by size and triple-expansion engines; not only could steamers now sail much further, but coal could be carried as cheaply in steamers as in sailing ships. The other occurrence was more than a matter of economics; it was a new attitude toward time. The industrial elites, imbued with a sense of commanding the destiny of the world, demanded of events both speed and predictability. Among sailing ships, only the clippers could compete with steamers in sheer speed, and they became deservedly famous for it. Not even the clippers, however, could meet a schedule. Once the thought had taken root that one could legitimately expect certain events to take place at certain precise times—an idea first implanted in people's minds by the experience of the railroads—it became increasingly unbearable to think that one's shipment or person might not arrive at a predictable moment. The qualms and doldrums and shifting winds of the open sea became the enemies of industrial man, creature of his calendars and clocks.[14]

In the maritime world, the spirit of scheduling was embodied in the institution of the shipping line. In 1875 a steamship kept on a regular schedule was the equivalent, in annual carrying capacity, of three sailing ships of comparable tonnage. By 1900 the ratio had risen to four-to-one.[15] Owners of expensive steamers had every incentive to use them to their maximum, for the investments were enormous, and obsolescence rapid. In other words, they had to schedule their customers to meet their timetables. And the customers were only too willing to cooperate, for goods-in-transit represented capital uselessly tied up.

The earliest steamship lines were heavily subsidized and regulated. The Peninsular and Oriental, for example, received £160,000 a year beginning in 1845 for its service to India and China. By the 1860s the British government was spending over a million pounds a year on mail contracts to steamship companies, several of which proudly bore the name "Royal Mail."[16]

In addition to carrying the mail, these contract steamers served the government as transports in wartime; in fact, they were built to Admiralty specifications for this purpose. And many did serve: P and O liners were requisitioned in the Burma campaign of 1852, in the Indian Mutiny of 1857, and in the Sudan expedition of 1884; ships of the Calcutta and Burmah Steam Navigation Co. served in 1857 and in the Abyssinian campaign of 1863; and Cunard liners were used in Cyprus in 1878, Egypt in 1882, and the Sudan in 1884–85.[17]

Like the P and O and the British India, lines to Africa also owed their beginnings to mail subsidies. Thus the Union Castle Line serving South Africa traces its origins to the Union Steam Collier Co., later called Union Steam Ship Co., which obtained the mail contract to Capetown in 1857. Between Britain and West Africa the pioneer line was the African Steamship Co. It was founded in 1851 by the Liverpool merchant Thomas Sterling and the indefatigable Macgregor Laird. Its purpose was to link Britain to the Niger basin via Laird's projected steamboat service on the Niger River, thus carrying palm oil and other products of the West African interior to Europe at producers' prices. This would have a double benefit. It would bypass the middlemen, both African and European, who hampered a beneficial trade by burdening it with outrageous markups. And it would deliver the perishable palm oil to Europe in a fresher condition by avoiding the notorious West African coastal doldrums that often delayed sailing ships for weeks on end. To start a steamship line, however, the new firm needed a subsidy. In 1852, Laird obtained a contract of £21,250 per year to carry the mail once a month between Britain and West Africa.

Yet there were limits on the amount of mail and the generosity of the British government. Steamships conquered the seas on their merits as safe, reliable, and speedy means of transporting freight. In the West African trade this happened in the 1860s, when the British and African Steam Navigation Co. began competing with the African Steamship Co. By 1880 three times more steamers than sailing ships entered Lagos, carrying six times more tonnage. In later years the competitors merged to form the Elder Dempster Line, which still serves the West African coast.[18]

In the Indian Ocean the introduction of unsubsidized shipping lines coincides with the three technological advances mentioned earlier: the compound engine, the Suez Canal, and the submarine cable. The cable was as important as the other two developments, for it allowed the headquarters of shipping companies to maintain contact with their ships, and shippers to coordinate their shipments with current schedules and the needs of their customers. It served also unscheduled, or "tramp," steamers, routing them according to the latest quotations on the world commodities market; in other words, it led them from places where they could buy low to places in which they could sell high. No longer were the economic fortunes of a ship entrusted to blind luck and the captain's business acumen. Whole fleets of ships were now controlled from headquarters, half a world away.

By the 1880s there were many British-owned steamship lines operating in the Indian Ocean and in Far Eastern waters. Though the first was the Peninsular and Oriental, the real pioneer among steamship lines was its rival the British India. This was the creation of a pious and energetic Scot, William Mackinnon. Born in Campbeltown, Argyleshire, in 1823, he began work as a clerk in Glasgow. In 1847 he went to India to seek his fortune and rose to a prominent position in one of India's leading trading houses, Smith, Mackenzie and Co. The Anglo-Indians of Calcutta, meanwhile, had become dissatisfied with the Peninsular and Oriental's monthly service and demanded better communications with the rest of the world. Mackinnon and several associates saw their opportunity, and in 1856 they founded the Calcutta and Burmah Steam Navigation Co. with three steamers. The following year they were awarded the Calcutta-Rangoon mail contract, and in 1862 the company was renamed the British India Steam Navigation Co. Mackinnon was its director and principal shareholder.

Mackinnon's thriving company soon introduced the compound engine into the Indian Ocean and later became the first to use the Suez Canal. Taking advantage of a booming market, Mackinnon extended his lines in all directions: to the Persian Gulf, to Singapore and Malacca, to the Dutch East Indies, to England, Australia, and China. By 1869 his fleet numbered fifty vessels, among them the most modern steamers available. While the P and O concentrated on the long-distance passenger and mail business, the British India became the most important cargo line of the Eastern seas. As Portuguese captains had discovered over four centuries earlier, there was as much profit carrying freight between the Eastern ports as to and from Europe. Thus by 1893 the British India Line had 110 ships covering twice as many miles as the Peninsular and Oriental routes.

Mackinnon was more than a successful shipping magnate. He was, as the *Times* put it, "a perfect type of old-fashioned Scotch Presbyterian," driven by his Calvinist morality to do his best in worldly and in moral affairs. In the 1870s he befriended Sir John Kirk, the British consul at Zanzibar, an enthusiast for the abolition of the slave trade in Africa. He also became friends with King Leopold II of Belgium and with General George "Chinese" Gordon. In 1872 the British India Line was awarded the mail contract between Aden, Zanzibar, and Natal. For Mackinnon this was the beginning of a deeper involvement in African affairs. By the 1880s he began losing interest in business and developed a spasmodic obsession with Africa. His faith in progress and Christian morality led him to believe that British rule would bring the blessings of civilization and salvation to those less fortunate. Hence he founded the Imperial British East Africa Co., chartered to extend the British Empire into East Africa. Like Macgregor Laird before him, Mackinnon combined the motives of the Crusaders with the methods of the Industrial Revolution.[19]

The British were not long alone in adopting the new shipping to link metropole and colonies. In the 1840s, French steamers between Marseilles and Alexandria proved so fast and luxurious that many British travelers preferred them, taking P and O ships only east of Suez. In the 1860s the Compagnie des Services Maritimes des Messageries Impériales (later changed to Messageries Maritimes) extended its operations from the Mediterranean to the Indian Ocean. A government-subsidized line like the P and O, it followed the French flag, first to Algeria, then to Indochina and China. In service and speed, it rivaled the Peninsular and Oriental, and as it also served India, Singapore, and Australia, it captured a considerable British clientele. By 1875 it owned almost two thirds of all French steam tonnage.[20]

The Messageries for the most part neglected France's African possessions. Beginning in 1856, a small paddle-wheeler, the *Guyenne,* stopped at Dakar on its way to and from South America. Consistent service, however, did not commence until the eighties under smaller firms such as Maurel et Prom (Bordeaux-Senegal) and Compagnie Fabre-Fraissinet (Marseilles-West Africa). Similarly, it was in the 1880s that a Belgian line, Walford et Cie., connected Belgium and the Congo, and that the first German steamship lines—the Woermann, the Atlas, and the Sloman—appeared in African ports. The lag of the French, Belgians, and Germans behind Britain cannot be explained so much by their delayed industrialization as by their delayed conquest of Africa and its economic insignificance in the nineteenth century, in comparison with India.[21]

Among empires, the most unusual kind is that of the sea. The Minoans, the Greeks, the Phoenicians, and the Vikings all dominated for a time the seas around them. But only once has there been a truly global thalassocracy, a nation whose fleet and merchant marine were dominant on almost all the seas of the world. This was Great Britain in the nineteenth century.

There were those who attributed the power of Britain to the unique virtues of the British national character, or to divine grace. Others—mostly foreigners—saw in it only proof of Albion's perfidy and greed; as one Arab sheik explained to a member of Chesney's Euphrates Expedition, "The English are like ants: if one finds a bit of meat, a hundred follow."[22] But thalassocracy requires something other than divine grace or perfidy; it requires a superior technology and an economy to back it up. In the early nineteenth century, Britain's hard-won maritime supremacy was threatened by an American ship-building boom based on limitless supplies of cheap timber. The shift to iron ships rescued Britain's dominance. From one quarter of the world's tonnage in 1840, Britain's share rose to 42.7 percent in 1850 and remained between forty and fifty percent until World War One.[23] Between 1890 and 1914, half the world's seaborne trade was carried in British-owned vessels, and Britain built two thirds of the world's new ships.[24]

Britain's lead in the steam-engine and metal-hull industries was reinforced by another factor. She possessed the richest deposits of the world's best steamer coal within a few miles of her coasts. Much of the coal later discovered near the Indian Ocean—in Natal, Bengal, and Borneo—was located in British colonies, as were the most convenient coaling stations. In fact Britain practically monopolized the world's naval coal supplies, just as she dominated the global submarine cable network.[25] (The importance of these factors is readily apparent when one contrasts them with their twentieth-century equivalents, petroleum and the radio, for neither of these gave any nation the edge that coal and cables had once given Britain.)

In the shadow of Britain's domination of the sea, lesser thalassocracies could also claim the title of empire. Some, like

France and Germany, were authentic great powers within Europe. The French, uncertain of their status after the defeat of 1871, found consolation in possessing more land than the British, while the Germans were half-proud and half-ashamed of the few backwater colonies they belatedly acquired. Nations that were small and weak by European standards—Italy, Portugal, Belgium, the Netherlands—used colonies to achieve imperial status. All in all, half the world was Europe Overseas, thalassocracies bound together by delicate webs of steamship lines and cables.[26]

The enormous expansion of shipping and communications in the nineteenth century was part of an even larger change in world economic relations. Throughout historic times and until the nineteenth century, trade between East and West bore two characteristics. One was the high cost of transportation, whether through the Middle East or around Africa, which made trade worthwhile only for the most precious goods: tea, porcelain, spices, indigo, silk, pearls, gems, and bullion. The other was the unbalanced nature of the trade, with the East offering products that Europeans coveted, while the Europeans had little to offer in return, save bullion and later opium. The steamships demolished this age-old system, and in its place fostered a new economic relationship in which both sides exchanged bulky commodities at low freight rates. Europe offered the products of its industry—cotton, machinery, iron, and coal—in exchange for Eastern raw materials and cheap agricultural goods: wheat, rice, jute, rubber, guttapercha, tin, and other products.[27]

Steam, cables, and the Suez Canal revolutionized the relations between East and West more than did the great discoveries five centuries earlier. The Portuguese caravels and Spanish galleons had closed the global ecumene, bringing peoples all around the world into contact with one another. The new technologies of the nineteenth century deepened these contacts into a constant flow of goods, people, and ideas. They turned isolated subsistence economies with limited trade contacts into parts of a single world market in basic commodities. They shattered traditional trade, technology, and political relationships, and in their place they laid the foundations for a new global civilization based on Western technology.

## NOTES

1. George Gibbard Jackson, *The Ship Under Steam* (New York, 1928), p. 149; W. A. Baker, *From Paddle-Steamer to Nuclear Ship: A History of the Engine-Powered Vessel* (London, 1965), p. 57.

2. A. Fraser-Macdonald, *Our Ocean Railways; or, the Rise, Progress, and Development of Ocean Steam Navigation* (London, 1893), p. 228, cites the Bessemer process, whereas Bernard Brodie, *Sea Power in the Machine Age: Major Naval Inventions and their Consequences on International Politics* (London, 1943), p. 164, emphasizes the Siemens-Martin method as the determining factor.

3. On steel ships, see Brodie, p. 164; Ambroise Victor Charles Colin, *La navigation commerciale au XIXe siècle* (Paris, 1901), pp. 57–58; Fraser-Macdonald, p. 228; Duncan Haws, *Ships and the Sea: A Chronological Review* (New York, 1975), pp. 163–64; Carl E. McDowell and Helen M. Gibbs, *Ocean Transportation* (New York, 1954), p. 29; "Ship," in *Encyclopaedia Britannica* (Chicago, 1973), 20:409–10; and René Augustin Verneaux, *L'industrie des transports maritimes au XIXe siècle et au commencement du XXe siècle*, 2 vols. (Paris, 1903), 2:10.

4. On the triple-expansion engine, see Colin, p. 51; Harold James Dyos and Derek Howard Aldcroft, *British Transport* (Leicester, 1969), p. 242; Fraser-Macdonald, pp. 210–20; Adam W. Kirkaldy, *British Shipping: Its History, Organisation and Importance* (London and New York, 1914), pp. 130–37; Thomas Main (M.E.), *The Progress of Marine Engineering from the Time of Watt until the Present Day* (New York, 1893), p. 70; Auguste Toussaint, *History of the Indian Ocean*, trans. June Guicharnaud (Chicago, 1966), p. 212; Verneaux, 2:39; and W. Woodruff, *Impact of Western Man: A Study of Europe's Role in the World Economy 1750–1960* (London, 1966), p. 239.

5. Colin, pp. 53 and 74–77; and Dyos and Aldcroft, p. 243.

6. Kirkaldy, appendix XVII.

7. Fraser-Macdonald, p. 102.

8. Dyos and Aldcroft, pp. 244–45; Halford Lancaster Hoskins, *British Routes to India* (London, 1928), p. 219; Woodruff, pp. 239–40; and Paul Bairoch, *Révolution industrielle et sous-développement*, 4th ed. (Paris, 1974), pp. 177–78.

9. Kirkaldy, p. 343 and appendix XIV.

10. E. A. Benians, "Finance, Trade and Communications, 1870–1895," in E. A. Benians, James Butler and C. E. Carrington, eds., *Cambridge History of the British Empire*, vol. 3: *The Empire-Commonwealth 1870–1919* (Cambridge, 1959), p. 201; Gerald S. Graham, "Imperial Finance, Trade and Communications 1895–1914," in *Cambridge History of the British Empire*, 3:457; Hoskins, p. 472; Toussaint, p. 213; and Verneaux, 1:314–15.

11. See the map in Kirkaldy; also W. E. Minchinton, "British Ports of Call in the Nineteenth Century," *Mariner's Mirror* 62 (May 1976):145–57.

12. See Gerald S. Graham, "The Ascendancy of the Sailing Ship, 1850–85," *Economic History Review* 9(1956):74–88.

13. Colin, pp. 4–5. On the demise of sailing ships, see also Benians, p. 203; Charles Ernest Fayle, *A Short History of the World's Shipping Industry* (London, 1933), pp. 244–46; Kirkaldy, p. 318; McDowell and Gibbs, p. 251; Toussaint, p. 211; Verneaux, 2:13–14; and Woodruff, pp. 242–43.

14. The epitome of schedule-mania, without a doubt, was the idea entertained by the German General Staff before 1914 that they could defeat a more powerful coalition of enemies by dint of split-second timing. Even Napoleon's faith in his lucky star seems less mad.

15. Colin, pp. 4–5.

16. For example, the Cunard Line was the nickname of the British and North American Royal Mail Steam Packet Co. See Roland Hobhouse Thornton, *British Shipping* (London, 1939), p. 40; Haws, pp. 119 and 133; and Fraser-Macdonald, p. 94.

17. Fraser-Macdonald, pp. 103, 112 and 121.

18. See P. N. Davies, *The Trade Makers: Elder Dempster in West Africa, 1852–1972* (London, n.d.), ch. 1; and his "The African Steam Ship Company," in John Raymond Harris, ed., *Liverpool and Merseyside: Essays in the Economic and Social History of the Port and its Hinterland* (Liverpool, 1969), pp. 212–38. See also Sir Alan Cuthbert Burns, *History of Nigeria*, 6th ed. (New York, 1963), pp. 263–

64; Anthony G. Hopkins, *An Economic History of West Africa* (New York, 1973), pp. 149–50; Christopher Lloyd, *The Search for the Niger* (London, 1973), p. 189; Murray, p. 29; and H. Moyse-Bartlett, *A History of the Merchant Navy* (London, 1937), p. 235.

19. On Mackinnon's career as an imperialist in East Africa, see John S. Galbraith, *Mackinnon and East Africa 1878–1895; a Study in the 'New Imperialism'* (Cambridge, 1972). On his maritime career, see Galbraith, pp. 1–45; Fraser-Macdonald, pp. 106–10; Haws, p. 149; Hoskins, pp. 412–20; and "Mackinnon, Sir William," in *Dictionary of National Biography*, 22:999.

20. Colin, pp. 168–69; Hoskins, pp. 263, 412–13, and 436; Toussaint, pp. 206–07; and Verneaux, 2:47–49.

21. On French and German lines to Africa, see Emile Baillet, "Le rôle de la marine de commerce dans l'implantation de la France en A.O.F.," *Revue Maritime* 135(July 1957):832–40; D. K. Fieldhouse, *Economics and Empire 1830–1914* (Ithaca, N.Y., 1973), p. 287; Hopkins, pp. 130 and 149; Roger Pasquier, "Le commerce de la Côte Occidentale d'Afrique de 1850 à 1870," in Michel Mollat, ed., *Les origines de la navigation à vapeur* (Paris, 1970), pp. 122–24; Verneaux, 1:322–33; and André Lederer, *Histoire de la navigation au Congo* (Tervuren, Belgium, 1965), p. 127.

22. W. F. Ainsworth, *Personal Narrative of the Euphrates Expedition*, 2 vols. (London, 1888), 2:197, quoted in Hoskins, p. 193.

23. These figures are from Kirkaldy (appendix XVII), who calculated one steamer-ton as the functional equivalent of four sailing-tons.

24. Dyos and Aldcroft, pp. 23–32; and Woodruff, p. 238.

25. Brodie, pp. 113–15; Minchinton, p. 151.

26. On the idea of thalassocracy, see Herbert Lüthy, "Colonization and the Making of Mankind," *Journal of Economic History* 21 (1961):483–95.

27. On the economics of colonial trade, see the books by Fieldhouse and Hopkins cited above.

# CHAPTER THIRTEEN

# The Railroads of India

But a more powerful agency than that of laws, roads, bridges, canals, or even education, was destined to arouse the Hindoo from his torpidity. The steam-engine . . . with its advance was overturning prejudices, uprooting habits, and changing customs as tenaciously held and dearly loved almost as life itself.

(Captain Edward Davidson, consulting engineer for railways to the government of Bengal)[1]

. . . the real operation, after all, is to make the Hindoos form the railways, and enable us to reap a large portion of the profits.

(Hyde Clark, lobbyist for Indian railroads)[2]

India has tempted European empire seekers since the Middle Ages. She was the prize that Columbus sought, that da Gama found, and that Portugal, Holland, France, and Britain fought over many times thereafter. Yet three centuries after da Gama, European influence was still predominantly coastal, weakening as it got further from Bombay, Madras, Calcutta, and other ports. The difficulties of inland transportation weighed heavily against the European presence.

India offered no formidable obstacles to penetration comparable to the African disease barrier. The Deccan is a large plateau that slopes down toward the Bay of Bengal and falls off abruptly on its western edge. The Gangetic Plain is equally accessible. Yet several hindrances did impede progress inland. During the dry season, dirt roads are open to pedestrians, bullock carts and pack animals; in the rainy summer monsoons, however, they become impassible. Except for the Ganges, the rivers of India are barely navigable in the dry season. Before the age of steam, the natural pathways of India were just adequate for slow travel and communication but closed to the economic transportation of most freight. Grain could not be taken more than fifty miles on bullock carts, for beyond that distance the animals ate more than they carried. Bales of cotton carried to Bombay arrived so dirty with road dust or so soaked with rain as to be almost worthless. Only precious cargoes of opium, indigo, and shellac could bear the cost of long-distance freight. India was a land of isolated villages and subsistence agriculture.

The 1840s were a time of railroad fever in the Western world, and most of all in Britain. Railroad enthusiasts dreamed of covering the whole earth with their iron rails and puffing clattering trains. And few parts of the world seemed so desperately in need of the new invention as Britain's prized colony. Building the railroad system of India became the most monumental project of the colonial era; it involved the largest international capital flow of the nineteenth century, and produced the fourth longest rail network on earth, behind only those of the United States, Canada and Russia. Today's India, too poor to afford automobiles or air travel for the masses, is probably the world's most railroad-dependent nation. Yet, for all its evident usefulness, this railroad network remains a cause of argument among historians, for it illustrates on a large scale the grandeurs and miseries of colonialist enterprise.

Several disparate interests coincided to create the Indian

railroads. There were the visionaries, the promoters, and the journalists, whose justifications for wanting railroads were often vaguely expressed; thus John Chapman, promoter of the Great Indian Peninsular Railway, wrote in 1850 of "the double hope of earning an honourable competency and of aiding in imparting to our fellow subjects in India, a participation in the advantages of the greatest invention of modern times."[3] The *Economist* noted in 1857 that railroads in India would spread "English arts, English men and English opinions," and the engineer Davidson regarded them as "firm and unaltered memorials of British rule."[4]

Others had a direct financial stake in railroads. The Anglo-Indian merchants and the East India trading houses in London saw railroads as a means of extending their business to inland towns. The P and O, newly arrived in India, took a great interest in a railroad to the coal mines of Raniganj, northwest of Calcutta. Manufacturers of cutlery, kitchenware, firearms, and sundry metal goods perceived a great potential market in the interior of the subcontinent. But most of all, it was the cotton barons of Lancashire who supported the Indian railroad projects. They had a double objective: to sell their products to the masses of Indians, and to secure a more reliable source of cotton than the United States. The American cotton failure of 1846 gave a great impetus to the railroad from Bombay to the cotton districts, and Chapman commented in 1848 that the merchants of Lancashire "consider it . . . as nothing more than an extension of their own line from Manchester to Liverpool."[5]

Finally, there were military motives. The strategic value of railroads is a familiar topic in the history of Germany, Russia, and the American West; it was a powerful motive in India as well. Lord Dalhousie, governor-general of India, wrote in a "Minute to the Court of Directors" in 1853 that railroads would provide "full intelligence of any event to be transmitted to Government at five times the speed now possible; as well as

the concentration of its military strength on every given point, in as many days as it would now require months to effect."[6] Dalhousie's prediction came true in the Indian Revolt of 1857, which led to a rash of railroad building in 1858 and 1859.[7]

The pioneer of the Indian railroad system was Rowland Macdonald Stephenson, a railroad engineer and a visionary who dreamed of laying tracks from Europe to India and China. His campaign was in many ways the continuation of the struggle to open steam communication, which had resulted in the victory of the P and O in 1842. In 1841 he tried to persuade the Court of Directors of the East India Company to subsidize railroads in India but was rebuffed. He then turned to lobbying. In the Calcutta newspaper *The Englishman* he presented his plan for a railroad network spreading outward from Calcutta, and justified it with the usual mix of business and security reasons. His plan met with the wholehearted approval of the Bengal government and the Calcutta merchants. With the support of some London businessmen, Stephenson then presented the Court of Directors with a new proposal for a railway company, this time boldly requesting a four percent guarantee on its profits. Though the Court of Directors rejected this plan also, they nonetheless formed the East Indian Railway Co. in March 1845. At the same time, another group headed by John Chapman formed the Great Indian Peninsula Railway, which planned to lay tracks radiating from Bombay. Later that year Stephenson went to India to survey the route from Calcutta to Delhi, and reported favorably on it to the stockholders of his company. Chapman meanwhile traveled to Bombay with the engineer G. T. Clark to survey the western half of the country.

All that was lacking was the approval of the East India Company, with a guarantee of profits. The promoters therefore organized a powerful publicity campaign, enlisting the

support of *The Times, The Economist,* and a number of bankers, businessmen, railroad engineers, and members of Parliament from Lancashire and the Midlands. Such was the influence of the railroad lobby that the East India Company surrendered, and in 1847 it granted the promoters a guarantee of five percent profit for twenty-five years, plus free land and other facilities. In 1849, Chapman's Great Indian Peninsula was authorized to build a line from Bombay to Kalyan, thirty-four miles away. Work began in February 1852, and in April 1853 the first twenty-four-mile section to Thana went into service.

Once the financial and political obstacles had been surmounted, several additional companies sprang into existence: the Madras Guaranteed (1852), the Bombay, Baroda, and Central India (1855), the Sind, Punjab, and Delhi (1856), the Eastern Bengal (1858), the Great Southern of India (1858), and the Calcutta and South-Eastern (1859). And all of them were given a guarantee of four-and-a-half or five percent profits, with any deficit to be paid by the Indian treasury, and any surplus to be shared equally by the railroad company and the treasury.[8]

It is this guarantee policy, rather than the railroads themselves, that has become the object of criticism, especially on the part of Indian historians. Critics point out that the guarantee promised the shareholders in Indian railroads—almost all of them wealthy Britons—that if their companies performed poorly, the taxpayers of India would take the loss, a "heads-I-win, tails-you-lose" proposition. Furthermore the guarantee has been accused of encouraging corruption, waste, and extravagance, since, as one critic put it, "it was immaterial to him [the investor] whether the funds that he lent were thrown into the Hooghly or converted into bricks and mortar."[9]

On the other hand, supporters of the guarantee argue that the capital invested in Indian railroads—some £95,000,000 between 1845 and 1875—would never have left Britain otherwise.

It turned what would have been a risky speculation in railroad construction into a gilt-edged investment in the solvency of the Indian government and in its ability to tax the Indian people. Thanks to the guarantee, the Indian railroads were able to raise money at a lower interest rate than almost any other foreign or colonial railroad before 1870. While it is true that the Indian treasury had to subsidize the railroads in their early years, this subsidy was justified by the social value of the railroads, which outweighed their financial returns.[10]

The guarantee problem aside, the fact remains that India obtained a well-built rail network at a reasonable cost. For the most part construction was easy because much of India is flat and labor was abundant. Two types of terrain, however, challenged the engineers. One was the Western Ghâts, where the edge of the Deccan plateau falls off to the coastal plain in a jagged precipice 1,800 to 2,000 feet high. James T. Berkley, chief engineer of the Great Indian Peninsula Railway, chose to confront the cliffs at two points, the Thull Ghât toward Delhi and the Bhore Ghât toward Madras. The Thull Ghât, which rose 972 feet in 9.326 miles, required thirteen tunnels and six viaducts; the Bhore Ghât, which climbed 1,831 feet in 15.86 miles, necessitated the construction of twenty-five tunnels and eight viaducts. On the steepest portions, Y-shaped reversing stations allowed trains to stop, switch their locomotives to the other end, and tackle the next section backwards. Construction of the ascents required unprecedented efforts. Up to 42,000 workers toiled at one time on the Bhore Ghât, while cutting and tunneling used up two and a half tons of gunpowder a day. The Bhore Ghât was finally finished in 1863 after seven years of rigorous labor, and the Thull Ghât in 1865.

The other natural obstacle was the great rivers of India which frequently brought devastating floods. To span them, engineers had to build bridges and approaches that dwarfed any in Europe. The Kishna River bridge, with its thirty-six 100-foot spans, was 3,855 feet long. The Jumna River had two bridges, one of ten 250-foot spans, the other of twelve. The conquests of the Ghâts and the rivers were among the most monumental achievements in railway building, a tribute to the labor of Indians and the skill and daring of British engineers.

Elsewhere there were few major difficulties. The East India Railway, which began in 1854 with a line from Calcutta to Raniganj on the Ganges, 121 miles away, reached Delhi, 1,130 miles distant, twelve years later. There it was met in 1870 by the Great Indian Peninsula line from Bombay. The following year the rails linked Bombay with Madras. By 1872, India had over 5,000 miles of track.

The first lines conformed to high technical standards, for the work was supervised by Indian Army engineers. The railroads were also built in one of the world's broadest gauges, 1.67 meters, or 5½ feet, which accommodated speeds and loads commensurate with the size of the subcontinent. And the cost was, all things considered, quite reasonable. Thanks to the support of the Indian government, the railroads incurred no legal expenses, received free land, and recruited cheap labor. The Indian railroads built up to 1868 cost, on the average, £18,000 per mile, compared with a cost of £62,000 per mile in Britain.[11]

The successful construction of the major trunk lines did not quell the criticism of the guarantee system. In 1869, Viceroy Lord Lawrence proposed to end the system in which, he said, "the whole profits go to the Companies, and the whole loss to the Government."[12] The Indian government thereupon undertook to build its own railroads, adding 2,175 miles to the network between 1869 and 1880. It was motivated in part by the need to bring food to famine areas, for in times of famine the traditional means of transporting food disappeared as the bullocks starved. To save money, many of the new lines were built to meter gauge, saddling India with two incompatible systems and prompting delays and pilferage of merchandise at the points of interconnection. So inefficient was state construction that in 1880 the state reverted to the guarantee system and began buying up existing private lines, signing contracts with the former owners to manage their former lines.

By 1902, British India (today's India, Pakistan, Bangladesh and Burma) had 25,936 miles of railroads, more than the rest of Asia put together, over three times as much as Africa, and more both in total and per capita than Japan. The major trunk lines, one third of the total mileage, were government-owned. The history of the Indian railroads illustrates the dangers of calling the nineteenth century an age of free enterprise. Empires were built and maintained by a mixed economy of state and private capitalism, a system designed to temper the efficiency and greed of the private sector with the inefficiency and social conscience of government.[13]

Railroads are more than tracks and trains; they are a whole new way of life, the forerunners of a new civilization. Their impact on Indian society was very different from that on its Western equivalent, because of the colonial context in which they were built. Let us cast a brief look at the consequences of railroads.

Many progressive-minded Westerners of the mid-nineteenth century believed railroads would bring the Industrial Revolution to India. Edward Davidson, the railroad engineer, declared that railroads

> . . . would surely and rapidly give rise within this empire to the same encouragement of enterprise, the same multiplication of produce, the same discovery of latent wealth, and to some similar progress in social improvement, that have marked the introduction of improved and extended communication in various kingdoms of the Western world.[14]

And Karl Marx prophesied in 1853:

> I know that the English millocracy intend to endow India with
> railways with the exclusive view of extracting at diminished ex-
> penses the Cotton and other raw materials for their manufac-
> tures. But when you have once introduced machinery into the
> locomotion of a country, which possesses iron and coals, you are
> unable to withhold it from its fabrication. You cannot maintain
> a net of railways over an immense country without introducing
> all those industrial processes necessary to meet the immediate
> and current wants of railway locomotion, and out of which
> there must grow the application of machinery to those branches
> of industry not immediately connected with railways. The rail-
> way system will therefore become, in India, truly the forerunner
> of modern industry. . . . Modern industry, resulting from the
> railway system, will dissolve the hereditary divisions of labor,
> upon which rest the Indian castes, those decisive impediments
> to Indian progress and Indian power.[15]

Alas, nothing of the sort happened. Almost all the private
capital spent on Indian railroads was raised in Britain; of the
50,000 holders of Indian railroad shares in 1868, only 400 were
Indian, because shares could only be traded in London. It was
the policy of the railroad companies, the East India Company,
and the British government to hire British contractors and dis-
courage Indian enterprises. Two fifths of the capital raised for
the railroads were spent in Britain. Skilled workers, foremen,
and engineers were brought from Britain and paid twice the
home rate, plus free passage, medical care, and allowances.
Rails, locomotives, rolling stock, and other iron goods were
imported. A lack of suitable timber for sleepers, resulting from
the unreliable practices of Indian timber merchants, led the
railroads to bring to India sleepers of Baltic fir creosoted in
England. Even British coal was sometimes preferred to the
cheaper Indian coal.[16]

The railroads substantially lowered freight costs in India;
this, indeed, had been one of Stephenson's major arguments
when he wrote to the Court of Directors of the East India
Company:

. . . the people of India are poor, and in many parts thinly scattered over extensive tracts of country; but on the other hand India abounds in valuable products, of a nature which are in a great measure deprived of a profitable market by want of a cheap and expeditious means of transport. It may therefore be assumed that remuneration for railroads in India must, for the present, be drawn chiefly from the conveyance of merchandise, and not from passengers.[17]

Woodruff calculated the cost of inland transport in India in United States cents per short-ton mile; in the 1830s, it was 12¢; in the 1860s, 8¢; in 1874, 2¢; and in 1900, .8¢.[18] Similarly, Paul Bairoch has noted a drop in land transport costs on the order of twenty-to-one, principally between 1860 and 1880.[19] This tremendous decline did not contribute, as Woodruff believed, to the industrialization of the subcontinent, but to its dependence on British industry. India exported raw cotton, jute, grain, and other agricultural products, and in return imported cotton cloth, metalware, and manufactured goods from Britain. In the process many of India's traditional handicrafts withered away. The craftsmen thus deprived of their employment began to flood the cities, where few new industries were growing to give work to the unemployed.[20] In 1853, Marx had predicted that "England has to fulfill a double mission in India: one destructive, the other regenerating—the annihilation of old Asiatic society, and the laying [of] the material foundations of Western society in Asia."[21] Half the program—the destructive mission—was accomplished; the other half had to await the end of British rule.

Contrary to Stephenson's prediction, within a few years passengers were the main source of revenue of the railroad companies. Liberated from nature's timeless restraints on human mobility, Indians flooded the cities and places of pilgrimage. Those who predicted that Brahmins, Untouchables, and members of other castes would refuse to sit together, proved also to be wrong. The only concession the railroads made to the caste system was to have trains stop for a few minutes at mealtimes, so passengers could get off and prepare their food in their own traditional manner.[22]

But if the railroads brought together Indians of different castes and regions, they demonstrated that a new caste system had arisen, one not blessed by age-old religious sanctions. On the railroads the workmen were Indian, but the best jobs—as stationmasters of large stations, drivers of express trains, and administrators—were held by Britons. The railroads deliberately lost money on their first-class passengers, who were for the most part British, and made it up by packing their third-class compartments with poor Indians. And in the stations, first-class passengers enjoyed waiting rooms and rooms "reserved for ladies," while third-class passengers had waiting sheds and rooms for "women only."[23] The railroads were indeed "firm and unaltered memorials of British rule." Was it a coincidence that Indian anti-British nationalism appeared at the same time as the railroads?[24]

# NOTES

1. Edward Davidson (Capt., R.E.), *The Railways of India: With an Account of their Rise, Progress and Construction, Written with the Aid of the Records of the India Office* (London, 1868), p. 3.

2. Quoted in Daniel Thorner, *Investment in Empire: British Railway and Steam Shipping Enterprise in India, 1825–1849* (Philadelphia, 1950), p. 12.

3. J. Chapman, "Letter to the Shareholders of the G.I.P.R." (London, 1850), quoted in W. J. Macpherson, "Investment in Indian Railways, 1845–72," *Economic History Review* 2nd series no. 7 (1955):182.

4. Macpherson, p. 177; Davidson, p. 4.

5. Quoted in Thorner, p. 96; see also pp. 8, 18–23, and 112–13, and Macpherson, pp. 182–85.

6. Quoted in Davidson, p. 87.

7. Macpherson, p. 179.

8. On the organization and financing of the first Indian railroads,

see Thorner, pp. 44–61, 126, 140–47, 169, and 177; Romesh Chunder Dutt, *The Economic History of India in the Victorian Age*, 3rd ed. (London, 1908), pp. 174–75; M. A. Rao, *Indian Railways* (New Delhi, 1975), pp. 14–20; and J. N. Westwood, *Railways of India* (Newton Abbot and North Pomfret, Vt., 1974), pp. 12–17.

9. On the antiguarantee side of the argument, see Dutt, pp. 353–56 and Rao, pp. 25–27.

10. This is the argument of Macpherson, pp. 177–81. Thorner (ch. 7) and Westwood (pp. 13–15 and 37) take a more balanced position.

11. On the construction of the Indian railroads up to 1871, see Frederick Arthur Ambrose Talbot, *Cassell's Railways of the World*, 3 vols. (London, 1924), 1:72–83 and 140–150; Davidson, pp. 231–78; Rao, pp. 20–23; and Westwood, pp. 18–35.

12. Rao, pp. 27–28.

13. On the railroads of India after 1871, see Rao, pp. 13–14, 24–31, and 268–69; Macpherson, p. 177; Dutt, pp. 246–50; Westwood, pp. 42–58; Carl E. McDowell and Helen M. Gibbs, *Ocean Transportation* (New York, 1954), pp. 37–38; and W. Woodruff, *Impact of Western Man: A Study of Europe's Role in the World Economy 1750–1960* (London, 1966), pp. 233 and 253.

14. Davidson, pp. 87–88.

15. Karl Marx, "The Future Results of British Rule in India," *New York Daily Tribune*, August 8, 1853, p. 5.

16. Westwood, pp. 31–37; Macpherson, p. 177; Rao, p. 14; Davidson, pp. 110–11; and John Bourne, C. E., *Indian River Navigation: A Report Addressed to the Committee of Gentlemen Formed for the Establishment of Improved Steam Navigation upon the Rivers of India, Illustrating the Practicality of Opening up Some Thousands of Miles of River Navigation in India, by the Use of a New Kind of Steam Vessel, Adapted to the Navigation of Shallow and Shifting Rivers* (London, 1849), pp. 26–27.

17. Quoted in Horace Bell, *Railway Policy in India* (London, 1894), pp. 3–4.

18. Woodruff, p. 254.

19. Paul Bairoch, *Révolution industrielle et sous-dévelopement*, 4th ed. (Paris, 1974), p. 179.

20. Woodruff, p. 233; Bairoch, pp. 173–87.

21. Marx, p. 5.

22. Westwood, pp. 23 and 71.

23. Westwood, pp. 72–73 and 81–82.

24. On this point, see Westwood, pp. 38–40.

CHAPTER FOURTEEN

# African Transportation: Dreams and Realities

To the colonialists of the nineteenth century, Africa needed new transportation systems at least as much as India. South of the Sahara Desert, they found the indigenous transportation systems wholly unsatisfactory. Canoes on the rivers were too small and slow. The tsetse fly barred pack animals, turning human beings into beasts of burden for both freight and white men. Not only was human porterage morally degrading (although not all Europeans disapproved of it), it was evil in practical ways as well. Porterage routes drained the surrounding countryside of able-bodied men, some being forcibly conscripted for the detested work, others fleeing from the recruiters. In underpopulated regions, porter caravans took both food and food-producing labor, leaving malnutrition in their wake. They spread syphilis, trypanosomiasis, smallpox, and other diseases. Finally, human porterage was inefficient, hence expensive.

So in Africa as in India, the colonialists' thoughts turned to railroads. The report of the Brussels Conference of 1876 recommended "the construction of railroads for the purpose of substituting economical and rapid means of transport for the present human porterage."[1] The railroad expert Balzer put the situation in quantitative terms. Porters, he pointed out, carried twenty-five to thirty kilograms (50–70 lbs.) a distance of twenty-five to thirty kilometers (15–20 mi.) a day, on the average. A train carrying fifty tons of freight at twenty kilometers per hour (13 mph) could thus do the work of 13,333 porters.[2]

The problem was to find places in Africa where the carrying capacity of 13,333 porters could be justified. Railroads need a heavy investment in infrastructure, and consequently require a considerable traffic to make them remunerative. Such traffic can consist either of passengers and processed goods—the traffic of cities—or of raw materials—the products of farms and mines.

Many of the railroads built in tropical Africa handled the second kind of traffic. There were groundnut and palm oil railroads in West Africa, copper railroads from Katanga and Rhodesia to the sea, and cotton railroads in the Sudan and Uganda. These lines, even when they were quite long (as from Katanga to South Africa) served essentially one purpose; they were feeder lines for the shipping companies that carried off to Europe the products of the African soil.

The other main *raison d'être* of railroads, to link population centers, was lacking in Africa before 1914, for there were few towns and no cities located inland. Outside of the mining and plantation districts, if railroads were built they would have to create their own demand, causing large populations to gather and cities to arise in places where there were only scattered villages. Thus the colonialists of Africa were caught between their own pressing transportation needs and the impossible economics of railroads. The result was a great many projects, and a few completed railroad lines, justified either by dreamy speculations of future profit, or by completely noneconomic reasons.

In 1873, Sir Garnet Wolseley proposed the construction of a narrow-gauge railroad in the Gold Coast from Cape Coast to the River Prah to transport troops and materiel for his Ashanti campaign.[3] Though this line was never built, military railroads were later completed by the British in the Sudan, by the Italians in East Africa, and, especially, by the French from Senegal toward the Niger valley. Additional lines were constructed in Uganda to hasten the end of the slave trade and in British West Africa to prevent the French from stealing the inland commerce. In every case railroads were seen as agents of development or, in the jargon of the time, of civilization. "Colonial railroads," wrote the German Balzer, "are the principal means to achieve the economic and political goals of a rational colonial policy of the mother country," and the Frenchman Lionel Wiener declared: "It is especially railroads that bring civilization behind them. . . ."[4]

The first railroad in Africa was the Alexandria-Cairo line built in the 1850s as part of the Red Sea Route to India. After that came a few short lines in South Africa in the 1850s and 60s; the big rush to build railroads in that part of the continent came in the 1870s after the discovery of diamonds in West Griqualand and in the 1880s when gold was found near Johannesburg. Before the turn of the century, South Africa, like Algeria, already had a respectable network.[5]

In tropical Africa the French were for a time the most enthusiastic railroad builders. In 1879, soon after beginning their penetration of the Western Sudan, they laid plans for a railroad from Senegal inland. Their first line was inaugurated in 1885 between Saint-Louis and Dakar, a distance of 163 miles. Another line, from Kayes on the Senegal River to Koulikoro on the upper Niger, was begun in 1881 and completed in 1906; this was primarily a military line whose purpose was to transport troops through unconquered territory. Yet another line, linking Konakry in French Guinea to the upper Niger, was built between 1899 and 1914, mostly for the export of natural rubber. After that the French did relatively little railroad building.[6]

The British were next to construct railroads in Africa. They built a line across Kenya from Mombasa to Lake Victoria in 1901. In West Africa they were slower than the French to claim vast stretches of the interior, and their one large colony, Nigeria, was well served by river steamers. There were financial considerations as well. After its Indian experience, the British government refused to guarantee railroad investments in West Africa. Nor would the home government simply give railroads to the colonies for political reasons, as the French were doing. Therefore each colony had to pay for its own railroads out of taxes, duties, and loans. Only after Joseph Chamberlain became colonial secretary in 1895 did construction begin in earnest.[7]

The Germans likewise got a late start. Their first lines went from Swakopmund to Windhoek in Southwest Africa (1897–1902) and from Tanga to Mombo in German East Africa (1894–1905). After 1904 they began a crash program of railroad building in both these colonies, in Cameroon, and in Togo.[8]

One independent African country, Ethiopia, also acquired a railroad in this period. In 1894, Emperor Menelik gave a concession to the Compagnie Impériale d'Ethiopie to build a line from Djibouti to the Nile; when this company went bankrupt in 1907, the work was continued by a French firm, the Compagnie du Chemin de fer Franco-Ethiopien de Djibouti à Addis-Abéba, which finished the work in 1918.[9]

By 1914, Africa already had the pattern of railroads that exists to this day. The northern and southern parts of the continent had fairly complete networks linking their major towns and agricultural and mining areas. In the period 1910–14, South Africa had 7,586 miles of railroads, followed by Algeria with 2,004 miles, Egypt with 1,485 miles, and Southwest Africa with 1,474 miles.[10] Railroad statistics from the thirties show more of the same: South Africa had 13,027 miles, Algeria had 3,007, Egypt had 2,799, the Belgian Congo had 2,064, and Southwest Africa had 1,680.[11] In per capita terms the contrasts are even stronger. Africa as a whole had 3.3 miles of railroad lines for every 10,000 inhabitants. Some regions had far less; for instance, French Equatorial Africa had only 1.06 miles, French West Africa had 1.55, and Egypt 2.05. In comparison South Africa had 18.4 miles and Southwest Africa, a land of many mines and few people, had 64.6 miles for every 10,000 inhabitants.[12]

In terms of railroads, the difference between Africa and India is astounding. India caught the railroad boom at its height and emerged from the colonial age with a complete and efficient railroad system. In contrast tropical Africa emerged from colonialism with only scattered unconnected lines, serving mainly Europe's need for raw materials. And now that trucks and aircraft provide the same (or better) service as railroads at a much lower initial outlay, the development of an African rail network is likely to be a long time coming.

To understand the needs and problems of modern transportation systems in Africa let us consider the opening of the Congo River basin. The Congo River (also called Zaïre and Lualaba) and its tributaries (the Ubangi, Sangha, Kasai, and others) form one of the largest networks of navigable rivers in the world, covering an area the size of Western Europe. The promise of such a network as a means of developing and exploiting central equatorial Africa was obvious to Europeans who entered the region. The great river basin, however, is not accessible from the sea, for its last few miles (between Kinshasa and Matadi) are broken by rapids and falls. This is the reason the Congo basin was not explored until some forty years after the penetration of the Niger.

Henry Stanley understood well the technological needs of the region he had crossed in 1875–77. In 1882 he declared that "the Congo basin was not worth a two-shilling piece in its present state. To reduce it into profitable order, a railroad must be made between the Lower Congo and the Upper

Congo, when with its accessibility will appear its value."[13] It took eight years—from 1890 to 1898—to build that railroad, against some of the most difficult terrain, climate, disease, and labor problems that ever beset such an enterprise. Meanwhile, Stanley and his employer, King Leopold II of Belgium, were eager to begin tapping the wealth of the Congo. This meant transporting steamboats in pieces on the heads of porters from Matadi to Stanley Pool on the upper Congo. The first of these boats, the little nine-ton *En Avant,* created enormous problems. Just to clear a path around the first falls between Vivi and Isangila, a distance of fifty-four miles, took an entire year, from February 1880 to February 1881. Not until December 1881 was the *En Avant* launched on the upper Congo. Once the road was cut and the workforce assembled, other steamers quickly followed: the Congo Free State's *Belgique, Espérance, A.I.A.,* and *Royal,* and the mission steamers *Peace* and *Henry Read.*[14] In 1887 alone, porters carried six steamers from Matadi to the new town of Leopoldville (today Kinshasa) on the Pool; between May and October of that year 60,000 men carried 992 tons of freight, most of it steamer parts. Steamers were, in fact, the backbone of the Congo Free State; as Leopold put it in 1886: "Take good care of our Marine; it about sums up all our governmental authority at this moment."[15]

By the time the first train arrived in Leopoldville in 1898, porters had carried forty-three steamers, weighing in all 865 tons, to the upper river. Of these, twenty-one belonged to the Congo Free State, six to the Belgian trading firm Société Anonyme Belge pour le Commerce du Haut Congo, five to the Dutch firm Nieuwe Afrikaansche Handelsvennootschap, three to the French Congo, and seven to various missionary societies.[16] These steamers, in addition to the railroad, absorbed ninety percent of the capital invested in the Congo between 1878 and 1898.[17]

The needs of the Congo basin attracted innovations as well as investments. By the 1880s sternwheelers, an American in-

vention, began replacing the British-style sidewheelers. At the end of the century, the screw-propeller, which had replaced the wheels on oceangoing ships fifty years earlier, made its appearance on Congo steamers. The difficulty had been to find a way to keep a propeller two feet or more in diameter entirely under water, yet no deeper than a few inches below the surface. The solution, developed by the English boat-builder Alfred Yarrow, was to place the propeller under the rear of the craft inside a tunnel which would fill with water as soon as the boat began to move. Finally, to facilitate the assembly of steamer kits and avoid the need to build launching ramps, Yarrow also created floating sections of boats that could be bolted together in the water; the first such boat was the thirty-ton *Stanley* which was shipped to the Congo in 1883.[18]

After the Matadi-Leopoldville railroad was completed in 1898, the transportation of steamers jumped. Among the first shipments carried on the new railroad was the *Brabant,* a steamer of 150 tons, five times the size of any other on the upper Congo at the time, and equipped with a dining room, a bathroom, and cabins for twenty-four passengers. By 1901 there were 103 steamers on the Congo river system, transporting Europeans and their machines upriver and natural rubber downriver.[19]

Building the Matadi-Leopoldville railroad necessitated a much more massive and sustained effort than that required for transporting steamers. Stanley and Leopold had discussed the railroad project in 1878, immediately following Stanley's return from his first trip to the Congo. At the time, the Congo was still an area open to the enterprise of all, and the first group to try and organize a railroad included Stanley, William Mackinnon, and several English entrepreneurs. When this plan fell through, the initiative passed to the Belgians. In 1887, Captain Albert Thys, aide-de-camp to King Leopold, founded the Compagnie du Congo pour le Commerce et l'Industrie, or C.C.C.I. Two years later this firm spawned the

Compagnie du Chemin de Fer du Congo, which was granted a concession to build the railroad.[20]

The distance to be covered between Matadi and Leopoldville was 241 miles. Construction lasted eight years, and cost three times more than the founders had expected. The terrain near Matadi, up the Palabala ramparts, was very rough; the first 5½ miles, which took two years to build, required 99 bridges, 1,250 viaducts, and gradients of up to twenty-to-one. Respiratory illnesses, hunger, dysentery, beri-beri, and malaria decimated the workers; 900 died in the first twenty-seven months, 1,800 altogether, not counting 132 European engineers and foremen. Up to 60,000 porters worked on the site at one time. As the area around the railroad was underpopulated, the company had to recruit workers at great distances. Some 2,000 workers were brought in from West Africa and Zanzibar, but many escaped, went on strike, or demanded to be repatriated. Finally, the British and French governments forbade the railroad to recruit any further in their African territories. The Belgians then brought in 300 Barbadians and 550 Chinese from Macao. The Barbadians rebelled, many were shot, and others died of disease. Of the Chinese, 300 died or ran off into the interior in the first few weeks, never to reappear.[21]

Despite the horrors of its construction, the railroad paid off handsomely, transporting the colonialists' equipment and exporting first natural rubber and later the copper of Katanga. In 1898–99 it carried 14,092 tons; in 1913–14, 87,082; and in 1919–20, 123,458. Profits averaged nine percent and reached thirteen percent in 1912–13. So great was the demand for transportation, in fact, that by World War One the railroad proved inadequate; plans were soon laid to replace it with a new, straighter, electrified broad-gauge line, with ten times the capacity of the old.[22]

Whatever judgment one casts upon the methods of its construction, at least the Matadi-Leopoldville railroad was necessary and useful. Other projects of the same period, in contrast, lie more in the realm of dreams than of practical business. The French were especially guilty of extravagant projects, for their colonialism was less business-oriented than that of the Belgians, the British, or the Dutch. Among their projects, few were as wasteful as those concerning the Sahara Desert.

Interest in the Sahara arose soon after the Franco-Prussian War. The first advocates of Saharan projects, as Henri Brunschwig has pointed out, were not imperialists so much as technocrats inspired by the recent completion of the Suez Canal and the Union Pacific Railroad: "What interested them most of all was neither mercantilism nor nationalism but the creative research of the inventor and builder."[23] In 1874, Captain E. Roudaire published a plan to create an inland sea in the *chotts,* or depressions, to the south of Tunisia by digging a canal to let the Mediterranean Sea flow into the desert. This idea stimulated discussions in the Paris Society of Geography and in the Academy of Sciences, and resulted in two missions of exploration in 1874 and 1876.[24]

The first to propose a railroad across the Sahara were the explorer Paul Soleillet and the engineer Adolphe Duponchel. In 1873, Soleillet led an expedition, financed by the Algiers Chamber of Commerce, to the Tuat Oasis. Three years later he and Duponchel publicized the idea of a railroad linking Algeria to the Niger. Duponchel attributed quasi-magical powers to such a railroad which, he said, would create "a vast colonial empire . . . a French India rivalling its British counterpart in wealth and prosperity."[25] The publicity encouraged Minister of the Navy and Colonies Admiral Jauréguiberry and Governor Brière de l'Isle of Senegal to begin construction of a railroad from Senegal toward the Niger valley, a project which was carried to completion in 1906.[26]

The idea also appealed to Minister of Public Works Charles de Freycinet, a graduate of the Ecole Polytechnique and an enthusiast for overseas expansion. In a report to President

Jules Grévy in 1879 he endorsed the railroad project for several reasons: France was closer to Africa than most European countries; the greatness of France demanded that she place herself at the head of the movement for the conquest of Africa; and the project was certainly feasible, for had the Americans not just recently built a railroad from New York to San Francisco?[27] Freycinet thereupon appointed a committee under Ferdinand de Lesseps "for the study of questions relating to the railroad communications of Algeria and Senegal with the Soudan." The National Assembly appropriated first 200,000 francs and later 600,000 for preliminary studies. In 1880 an exploring party led by Colonel Flatters left Algeria in the direction of Lake Chad, but was massacred by Tuaregs under Turkish instigation. This proved that building a railroad through unconquered territory would be impossible, and the project was shelved for a decade.[28]

The Trans-Saharan idea reappeared in 1888–90 after France had penetrated from Senegal to the Niger and from Gabon to the Congo. As time passed, projects became more elaborate and fanciful. Two railroads were proposed, one to link Algeria with Lake Chad, the other with Senegal or Dahomey. Other projects involved railroads from Algiers to Djibouti and from Lake Chad to Johannesburg, for the convenience of British travelers in a hurry.[29] One retired cavalry officer suggested that a canal be dug from Timbuktu on the Niger to the Atlantic coast near the Canary Islands.[30] And a railroad expert, writing around 1910, looked forward to an even more grandiose future:

> By the year 1925 there is every reason to believe we shall have a complete and fairly direct Trans-African Railway from the extreme north to the extreme south, and we shall be able to traverse Africa by a variety of more or less meandering routes from west to east.[31]

But these projects all encountered the same obstacle. The cost would have been prohibitive, for French Africa was not another India, and the Trans-Saharan could never have paid for itself. Today a line reaches 160 miles into the Sahara to Colomb-Béchar, while columns of trucks roar across the desert, and airplanes fly overhead.[32]

## NOTES

1. André Lederer, *Histoire de la navigation au Congo* (Tervuren, Belgium, 1965), p. 129. The same recommendation was repeated in 1889–90; see Olufemi Omosini, "Railway Projects and British Attitudes Toward the Development of West Africa, 1872–1903," *Journal of the Historical Society of Nigeria* 5(1971):502.

2. F. Balzer, *Die Kolonialbahnen, mit besonderer Berücksichtigung Afrikas* (Berlin and Leipzig, 1916), pp. 21–22.

3. Omosini, p. 493.

4. Balzer, pp. 15–16; Lionel Wiener, *Les chemins de fer coloniaux de l'Afrique* (Brussels and Paris, 1931), p. 5.

5. Wiener, pp. 20–21 and 340–42.

6. Emile Baillet, "Le rôle de la marine de commerce dans l'implantation de la France en A.O.F.," *Revue Maritime* 135(July 1957): 837; Wiener, pp. 82–93; and Jacques Mangolte, "Le chemin de fer de Konakry au Niger (1890–1914)," *Revue française d'histoire d'outre-mer* 55(1968):37–105.

7. Omosini, pp. 492–506.

8. Balzer, pp. 27–29.

9. Wiener, pp. 134–37.

10. Balzer, pp. 98–99 and 463.

11. Col. J. Mornet, "L'outillage comparé des différents pays d'Afrique," *L'Afrique Française, Bulletin Mensuel du Comité de l'Afrique Française et du Comité du Maroc* 44 no. 10 (Oct. 1934): 580–84; see also Wiener, pp. 142 and 562–63.

12. Mornet, p. 581.

13. Henry Morton Stanley, *In Darkest Africa, or the Quest, Rescue, and Retreat of Emin, Governor of Equatoria,* 2 vols. (New York, 1890), 1:463.

14. Lederer, pp. 39 and 56–58.

15. Lederer, p. 95.

16. Lederer, p. 130.

17. André Huybrechts, *Transports et structures de développement au Congo. Etude du progrès économique de 1900 à 1970* (Paris, 1970), pp. 9–10.

18. Eleanor C. Barnes, *Alfred Yarrow: His Life and Work* (London, 1924), ch. 14 and p. 164; and Alfred F. Yarrow, "The Screw as a Means of Propulsion for Shallow Draught Vessels," *Transactions of the Institution of Naval Architects* 45(1903):106–17.

19. Lederer, pp. 111 and 134–37.

20. René-Jules Cornet, *La bataille du rail. La construction du chemin de fer de Matadi au Stanley Pool* (Brussels, 1947), pp. 29–81; Wiener, pp. 190–93; Frederick Arthur Ambrose Talbot, *Cassell's Railways of the World*, 3 vols. (London, 1924), 3:610–16.

21. Huybrechts, pp. 11–12; Wiener, pp. 195–97; Cornet, *Bataille*, pp. 125–341.

22. Huybrechts, p. 71; Balzer, pp. 23–24; and Wiener, pp. 193–94.

23. Henri Brunschwig, "Note sur les technocrates de l'impérialisme français en Afrique noire," *Revue française d'histoire d'outre-mer* 54(1967):171–73.

24. Captain E. Roudaire, "Une mer intérieure en Algérie," *Revue des Deux Mondes* 3(May 1874):323–50; Agnes Murphy, *The Ideology of French Imperialism 1871–1881* (Washington, 1948), pp. 70–75; and Brunschwig, "Note," pp. 173–75.

25. Adolphe Duponchel, *Le Chemin de fer transsaharien, jonction coloniale entre l'Algérie et le Soudan* (Montpellier, 1878), p. 218, quoted in Alexander S. Kanya–Forstner, *The Conquest of the Western Sudan: A Study in French Military Imperialism* (Cambridge, 1969), p. 61.

26. Omosini, p. 499; Archives Nationales Section Outre-Mer (Paris), Afrique XII, Dossier 2a: various papers concerning the Senegal-Niger railroad project; and Kanya–Forstner, pp. 64–69.

27. Archives Nationales Section Outre-Mer (Paris): Afrique XII, Dossier 2a no. 455, July 12, 1879.

28. Archives, Dossier 2b; Kanya–Forstner, pp. 62–67; Murphy, pp. 85–90; and Brunschwig, "Note," pp. 176–78.

29. Brunschwig, "Note," pp. 176–83.

30. Archives Nationales Section Outre-Mer (Paris): Afrique XII, Dossier 4: "Project Levasseur de Tombouctou à la Mer," and Travaux Publics, A.O.F., Carton 47, Dossier 3: "Canal maritime de Tombouctou à la mer. Proposition Levasseur 1896."

31. Ernest Protheroe, *The Railways of the World* (London, n.d.), p. 652.

32. On later projects, see Paul Leroy-Beaulieu, *Le Sahara, le Soudan et les chemins de fer transsahariens* (Paris, 1904); Commandant Roumens, *L'impérialisme français et les chemins de fer transafricains* (Paris, 1914); and Brunschwig, "Note," pp. 184–86.

CHAPTER FIFTEEN

# The Legacy of
# Technological Imperialism

The history of European imperialism in the nineteenth century still contains a number of paradoxes, which an understanding of technology can help elucidate. One of them is the expansion of Britain in the mid-century, a world power claiming to want no more imperial responsibilities yet reluctantly acquiring territories "in a fit of absent-mindedness." Was this really a case, as Fieldhouse put it, of "a metropolitan dog being wagged by its colonial tail"? A more appropriate metaphor might be the pseudonym Macgregor Laird used in writing to *The Spectator:* Cerberus, the many-headed dog.

For the imperialist drive did not originate from only one source. In the outposts of empire, and most of all in Calcutta and Bombay, were eager imperialists, adventurous and greedy for territory. They lacked, however, the industry to manufacture the tools of conquest. Had they been able to create the instruments appropriate to their ambitions, they might well have struck out on their own, like the settlers in the Thirteen Colonies of North America. But against Burma, China, the Middle East, and Africa they needed British technology.

In Britain, meanwhile, the politicians were at times reluctant; the lengthy delay in occupying Egypt is an example of this. But the creators of the tools of empire—people like Peacock, the Lairds, the arms manufacturers—were provisioning the empire with the equipment that the peripheral imperialists required. The result was a secondary imperialism, the expansion of British India, sanctioned, after the fact, by London.

Imperialism in the mid-century was predominantly a matter of British tentacles reaching out from India toward Burma, China, Malaya, Afghanistan, Mesopotamia, and the Red Sea. Territorially, at least, a much more impressive demonstration of the new imperialism was the scramble for Africa in the last decades of the century. Historians generally agree that from a profit-making point of view, the scramble was a dubious undertaking. Here also, technology helps explain events.

Inventions are most easily described one by one, each in its own technological and socioeconomic setting. Yet the inner logic of innovations must not blind us to the patterns of chronological coincidence. Though advances occurred in every period, many of the innovations that proved useful to the imperialists of the scramble first had an impact in the two decades from 1860 to 1880. These were the years in which quinine prophylaxis made Africa safer for Europeans; quick-firing breech-loaders replaced muzzle-loaders among the forces stationed on the imperial frontiers; and the compound engine, the Suez Canal, and the submarine cable made steamships competitive with sailing ships, not only on government-subsidized mail routes, but for ordinary freight on distant seas as well. Europeans who set out to conquer new lands in 1880 had far more power over nature and over the people they encountered than their predecessors twenty years earlier had; they could accomplish their tasks with far greater safety and comfort.

Few of the inventions that affected the course of empire in the nineteenth century were indispensable; quinine prophylaxis comes closest, for it is unlikely that many Europeans would willingly have run the risks of Africa without it. The muzzle-loaders the French used in fighting Abd-el Kader could also have defeated other non-Western peoples; but it is unlikely that any European nation would have sacrificed for Burma, the Sudan, or the Congo as much as France did for Algeria.

Today we are accustomed to important innovations being so complex—computers, jet aircraft, satellites, and weapons systems are but a few examples—that only the governments of major powers can defray their research and development costs; and generally they are eager to do so. In the nineteenth century European governments were preoccupied with many things other than imperialism. Industrialization, social conflicts, international tensions, military preparedness, and the striving for a balanced budget all competed for their attention. Within the ruling circles of Britain, France, Belgium, and Germany, debates raged on the need for colonies and the costs of imperialism.

What the breechloader, the machine gun, the steamboat and steamship, and quinine and other innovations did was to lower the cost, in both financial and human terms, of penetrating, conquering, and exploiting new territories. So cost-effective did they make imperialism that not only national governments but lesser groups as well could now play a part in it. The Bombay Presidency opened the Red Sea Route; the Royal Niger Company conquered the Caliphate of Sokoto; even individuals like Macgregor Laird, William Mackinnon, Henry Stanley, and Cecil Rhodes could precipitate events and stake out claims to vast territories which later became parts of empires. It is because the flow of new technologies in the nineteenth century made imperialism so cheap that it reached the threshold of acceptance among the peoples and governments of Europe, and led nations to become empires. Is this not as important a factor in the scramble for Africa as the political, diplomatic, and business motives that historians have stressed?

All this only begs a further question. Why were these innovations developed, and why were they applied where they
would prove useful to imperialists? Technological innovations
in the nineteenth century are usually described in the context
of the Industrial Revolution. Iron shipbuilding was part of
the growing use of iron in all areas of engineering; submarine
cables resulted from the needs of business and the development of the electrical industry. Yet while we can (indeed, we
must) explain the invention and manufacture of specific new
technologies in the context of general industrialization, it does
not suffice to explain the transfer and application of these
technologies to Asia and Africa. To understand the diffusion
of new technologies, we must consider also the flow of information in the nineteenth century among both Western and non-
Western peoples.

In certain parts of Africa, people are able to communicate
by "talking drums" which imitate the tones of the human
voice. Europeans inflated this phenomenon into a great myth,
that Africans could speak to one another across their continent by the throbbing of tom-toms in the night. This myth of
course reflected the Westerners' obsession with long-range communication. In fact, nineteenth-century Africans and Asians
were quite isolated from one another and ignorant of what
was happening in other parts of the world. Before the Opium
War, the court of the Chinese emperor was misinformed about
events in Canton and ignorant of the ominous developments
in Britain, Burma, and Nigeria. People living along the Niger
did not know where the river came from, nor where it went.
Stanley encountered people in the Congo who had never before heard of firearms or white men. Throughout Africa, warriors learned from their own experiences but rarely from those
of their neighbors.

To be sure, there were cases in which Africans or Asians
adopted new technologies. Indian princes hired Europeans to
train their troops. The Ethiopian Bezbiz Kasa had an English

sergeant make cannons for him, while Samori Touré sent a blacksmith to learn gunsmithing from the French. Mehemet Ali surrounded himself with European engineers and officers in a crash program to modernize his country. What is remarkable about these efforts is their rarity and, in most cases, their insufficiency. In the nineteenth century, only Japan succeeded in keeping abreast of Western technological developments.

In contrast, Western peoples—whether Europeans or descendants of Europeans settled on other continents—were intensely interested in events elsewhere, technological as well as otherwise. Physicians in Africa published their findings in France and Britain. American gun manufacturers exhibited their wares in London, British experts traveled to America to study gunmaking, and General Wolseley paid a visit to the American inventor Hiram Maxim to offer suggestions. Macgregor Laird was inspired by news of events on the Niger to try out a new kind of ship. Dutch and British botanists journeyed to South America to obtain plants to be grown in Asia. Scientists in Indonesia published a journal in French and German for an international readership. The latest rifles were copied in every country and sent to the colonies for testing. The mails and cables transmitted to and from the financial centers of Europe up-to-date information on products, prices, and quantities of goods around the world. And the major newspapers, especially the London *Times*, sent out foreign correspondents and published detailed articles about events in faraway lands. Then, as now, people in the Western world were hungry for the latest news and interested in useful technological innovations. Thus what seemed to work in one place, whether iron river steamers, quinine prophylaxis, machine guns, or compound engines, was quickly known and applied in other places. In every part of the world, Europeans were more knowledgeable about events on other continents than indigenous peoples were about their neighbors. It is the Europeans who had the "talking drums."

European empires of the nineteenth century were economy empires, cheaply obtained by taking advantage of new technologies, and, when the cost of keeping them rose a century later, quickly discarded. In the process, they unbalanced world relations, overturned ancient ways of life, and opened the way for a new global civilization.

The impact of this technologically based imperialism on the European nations who engaged in it is still hotly debated. The late nineteenth and early twentieth centuries were a time of overweening national pride, of frantic, often joyful, preparations for war. The cheap victories on the imperial frontiers, the awesome power so suddenly acquired over the forces of nature and over whole kingdoms and races, were hard to reconcile with the prudence and compromises which the delicate European balance required.

The era of the new imperialism was also the age in which racism reached its zenith. Europeans, once respectful of some non-Western peoples—especially the Chinese—began to confuse levels of technology with levels of culture in general, and finally with biological capacity. Easy conquests had warped the judgment of even the scientific elites.


Among Africans and Asians the legacy of imperialism reflects their assessment of the true value of the civilization that conquered them. Christianity has had little impact in Asia, and its spread in Africa has been overshadowed by that of Islam. Capitalism, that supposed bedrock of Western civilization, has failed to take root in most Third World countries. European concepts of freedom and the rule of law have fared far worse. The mechanical power of the West has not brought, as Macgregor Laird had hoped, "the glad tidings of 'peace and good will toward men' into the dark places of the earth which are now filled with cruelty."

The technological means the imperialists used to create

their empires, however, have left a far deeper imprint than the ideas that motivated them. In their brief domination, the Europeans passed on to the peoples of Asia and Africa their own fascination with machinery and innovation. This has been the true legacy of imperialism.

# Bibliographical Essay

The information contained in this book came from hundreds of sources, most of them cited in the footnotes. Of the published sources, a few dozen were especially helpful, and I recommend them to the reader wishing to pursue certain topics in greater detail. For a general introduction to the theme, see Daniel R. Headrick, "The Tools of Imperialism: Technology and the Expansion of European Colonial Empires in the Nineteenth Century," *Journal of Modern History* 51 no. 2 (June 1979):231–63.

PART ONE

## STEAMBOATS AND QUININE, TOOLS OF PENETRATION

For the early history of steamboats in Asia, see Henry T. Bernstein, *Steamboats on the Ganges: An Exploration in the History of India's Modernization through Science and Technology* (Bombay, 1960), a model monograph in the social history of technology; H. A. Gibson-Hill, "The Steamers Employed in Asian Waters, 1819–39," *The Journal of the Royal Asiatic Society, Malayan Branch* 27 pt. 1 (May 1954):127–61; and Gerald S. Graham, *Great Britain in the Indian Ocean: A Study of Maritime Enterprise, 1810–1850* (Oxford, 1968).

The most readable recent account of the Opium War is Peter Ward Fay, *The Opium War, 1840–1842: Barbarians in the Celestial*

*Empire in the Early Part of the Nineteenth Century and the War by which They Forced Her Gates Ajar* (Chapel Hill, N.C., 1975). On the steamers employed in that war, see William Dallas Bernard, *Narrative of the Voyages and Services of the Nemesis from 1840 to 1843*, 2 vols. (London, 1844), but note that there is a second edition of this work (London, 1845) and a third edition: Captain William H. Hall (R.N.) and William Dallas Bernard, *The Nemesis in China, Comprising a History of the Late War in that Country, with a Complete Account of the Colony of Hong Kong* (London, 1846). See also Gerald S. Graham, *The China Station: War and Diplomacy 1830–1860* (Oxford, 1978).

Steamers in the European penetration of Africa are best described in Macgregor Laird and R. A. K. Oldfield, *Narrative of an Expedition into the Interior of Africa, by the River Niger, in the Steam-Vessels Quorra and Alburkah, in 1832, 1833, and 1834*, 2 vols. (London, 1837); Christopher Lloyd, *The Search for the Niger* (London, 1973); and André Lederer, *Histoire de la navigation au Congo* (Tervuren, Belgium, 1965).

Still the best biography of Peacock is Carl Van Doren, *The Life of Thomas Love Peacock* (London and New York, 1911).

Malaria and quinine prophylaxis in the penetration of Africa are dealt with in Philip D. Curtin, *The Image of Africa: British Ideas and Actions 1780–1850* (Madison, Wis., 1964), a brilliant piece of historical research; but see also Michael Gelfand, *Rivers of Death in Africa* (London, 1964).

PART TWO

# GUNS AND CONQUESTS

There is a bewildering choice of books on guns and rifles. The ones I found most useful were William Young Carman, *A History of Firearms from Earliest Times to 1914* (London, 1955); William Wellington Greener, *The Gun and Its Development; With Notes on Shooting*, 9th ed. (London, 1910); Graham Seton Hutchison, *Machine Guns, Their History and Tactical Employment (Being Also a History of the Machine Gun Corps, 1916–1922)* (London, 1938); and H. Ommundsen and Ernest H. Robinson, *Rifles and Ammunition* (London, 1915).

On the use of firearms in the colonial wars of the nineteenth century, especially in Africa, see a series of interesting articles in the *Journal of African History* vols. 12 (1971) and 13 (1972); and also

Brian Bond, ed., *Victorian Military Campaigns* (London, 1967); Col. Charles E. Callwell, *Small Wars: Their Principles and Practice,* 3rd ed. (London, 1906); and Michael Crowder, ed., *West African Resistance: The Military Response to Colonial Occupation* (London, 1971).

PART THREE

# THE COMMUNICATIONS, REVOLUTION

The history of steamships and shipping in the nineteenth century, like that of firearms, has received a great deal of attention. I have found the following works most helpful: Ambroise Victor Charles Colin, *La navigation commerciale au XIXe siècle* (Paris, 1901); A. Fraser-Macdonald, *Our Ocean Railways; or, the Rise, Progress, and Development of Ocean Steam Navigation* (London, 1893); Adam W. Kirkaldy, *British Shipping: Its History, Organisation and Importance* (London and New York, 1914); Carl E. McDowell and Helen M. Gibbs, *Ocean Transportation* (New York, 1954); Michel Mollat, ed., *Les origines de la navigation à vapeur* (Paris, 1970); Roland Hobhouse Thornton, *British Shipping* (London, 1939); and René Augustin Verneaux, *L'industrie des transports maritimes au XIXe siècle et au commencement du XXe siècle,* 2 vols. (Paris, 1903).

On communications with overseas colonies, see Frank James Brown, *The Cable and Wireless Communications of the World; a Survey of Present Day Means of International Communication by Cable and Wireless, Containing Chapters on Cable and Wireless Finance* (London and New York, 1927); Bernard S. Finn, *Submarine Telegraphy: The Grand Victorian Technology* (Margate, 1973); Halford Lancaster Hoskins, *British Routes to India* (London, 1928, reprinted 1966), still the definitive study on this subject; P. M. Kennedy, "Imperial Cable Communications and Strategy, 1870–1914," *English Historical Review* 86(1971):728–52; John Marlowe, *World Ditch: The Making of the Suez Canal* (New York, 1964); and René Augustin Verneaux, *L'industrie des transports maritimes au XIXe siècle et au commencement du XXe siècle* (Paris, 1903).

On the railroads of India and Africa see M. A. Rao, *Indian Railways* (New Delhi, 1975); Daniel Thorner, *Investment in Empire: British Railway and Steam Shipping Enterprise in India, 1825–1849* (Philadelphia, 1950); J. N. Westwood, *Railways of India* (Newton Abbot and North Pomfret, Vt., 1974); and Lionel Wiener, *Les chemins de fer coloniaux de l'Afrique* (Brussels and Paris, 1931).

# Index